Unravelling Molecular Docking - From Theory to Practice

Edited by Črtomir Podlipnik

Published in London, United Kingdom

Unravelling Molecular Docking - From Theory to Practice
http://dx.doi.org/10.5772/intechopen.111324
Edited by Črtomir Podlipnik

Contributors
Ajay Kumar Shukla, Daling Zhu, Hang Yu, Irshad Begum, Ismail Celik, Joohee Pradhan, Krupa Mukeshbhai Santoki, Merve Yildirim, Mohammad Ahmad, Mohd Mursal, Mohemmed Faraz Khan, Muhammad Imran, Muktesh Chandra, Nandini Navinchandra Vadhavana, Neha Mathur, Noopur Khare, Phuc-Chau Do, Pragati Khare, Rabia Bushra, Rabya Munawar, Sahil Hussain, Sana Shamim, Sarita, Sejal Shah, Shankar K.M., Sheikh Muhammad Zesshan Qadar, Shine P. Varkey, Shristhi Shandily, Siva Sai Chandragiri, Tehseen Quds, Tu H.T. Nguyen, Vy T.T. Le, Yasmeen Rashid, Yingqi Wang

First published in London, United Kingdom, 2025 by IntechOpen
IntechOpen is the global imprint of INTECHOPEN LIMITED, registered in England and Wales, registration number: 11086078, 167-169 Great Portland Street, London, W1W 5PF, United Kingdom

For EU product safety concerns: IN TECH d.o.o., Prolaz Marije Krucifikse Kozulić 3, 51000 Rijeka, Croatia, info@intechopen.com or visit our website at intechopen.com.

British Library Cataloguing-in-Publication Data
A catalogue record for this book is available from the British Library

Unravelling Molecular Docking - From Theory to Practice
Edited by Črtomir Podlipnik
p. cm.

This title is part of the Biomedical Engineering Book Series, Volume 27
Topic: Bioinformatics and Medical Informatics
Series Editor: Robert Koprowski
Topic Editor: Slawomir Wilczynski

Print ISBN 978-0-85466-842-7
Online ISBN 978-0-85466-841-0
eBook (PDF) ISBN 978-0-85466-843-4
ISSN 2631-5343

IntechOpen Book Series
Biomedical Engineering
Volume 27

Aims and Scope of the Series

Biomedical Engineering is one of the fastest-growing interdisciplinary branches of science and industry. The combination of electronics and computer science with biology and medicine has improved patient diagnosis, reduced rehabilitation time, and helped to facilitate a better quality of life. Nowadays, all medical imaging devices, medical instruments, or new laboratory techniques result from the cooperation of specialists in various fields. The series of Biomedical Engineering books covers such areas of knowledge as chemistry, physics, electronics, medicine, and biology. This series is intended for doctors, engineers, and scientists involved in biomedical engineering or those wanting to start working in this field.

Meet the Series Editor

Robert Koprowski, MD (1997), Ph.D. (2003), Habilitation (2015), is an employee of the University of Silesia, Poland, Institute of Computer Science, Department of Biomedical Computer Systems. For 20 years, he has studied the analysis and processing of biomedical images, emphasizing the full automation of measurement for a large inter-individual variability of patients. Dr. Koprowski has authored more than a hundred research papers with dozens in impact factor (IF) journals and has authored or co-authored six books. Additionally, he is the author of several national and international patents in the field of biomedical devices and imaging. Since 2011, he has been a reviewer of grants and projects (including EU projects) in biomedical engineering.

Meet the Volume Editor

Dr. Črtomir Podlipnik obtained his Ph.D. in theoretical chemistry at the University of Ljubljana under Prof. Jože Koller. He later joined Prof. Anna Bernardi's group at the University of Milan, working on cholera toxin inhibitors. Now an assistant professor at the University of Ljubljana, his research focuses on quantum chemical computing, chemoinformatics, and drug design, including SARS-CoV-2 and Ebola inhibitors. He is involved in the ERASMUS+ Mundus Chemoinformatics+ program and the ECTN Virtual Education Committee, contributing to digital education and computational chemistry. He actively participates in international projects on biomolecular research.

Contents

Preface

Molecular docking is a crucial computational tool that forms the backbone of today's drug discovery, structural biology and the study of biomolecular contacts. Computer simulation of molecule-molecule binding aids in molecular recognition, prediction of binding affinities, and development of new therapies. This book provides a concise and clear overview of molecular docking in two sections: "Methods" and "Applications".

The "Methods" section covers important fundamentals and computational approaches in molecular docking. It covers docking algorithms, scoring functions and optimizations with respect to important parameters for sound prediction. It mentions rigid and flexible docking approaches, ensemble docking approaches and the integration of molecular dynamics simulations to improve docking accuracy and confidence. Ligand and receptor flexibility, solvation, and artificial intelligence in improving the accuracy of docking predictions, including benchmarking approaches and best practices for reproducing and validating docking studies, are discussed in detail in this section.

The "Applications" section looks at the impact of molecular docking in a range of research areas. Here, docking in drug discovery is discussed, starting from virtual compound library screening to lead optimization and drug repurposing. Molecular docking will be addressed in the study of protein-protein and protein-nucleic acid interfaces and their use in the conventional docking of small molecules. Case studies will be used to illustrate the contribution of docking in the discovery of therapeutic candidates for treating infectious diseases, malignancies and genetic diseases. The integration of docking into personalized therapy approaches and its future in the customization of drugs for individual patients is discussed in detail.

The book is a practical resource for students, professionals and researchers involved in molecular docking studies as it provides both theoretical background and practice-oriented perspectives. The detailed discussion of methods contributes to a deeper understanding of computational methods, and the real-world information illustrates the translational value of docking in the biomedical and pharmaceutical sciences.

We want to thank all the authors who contributed their expertise to the compilation of this volume. Their combined efforts have created a valuable reference work that combines computational methods with practical applications. We would also like to thank the editorial team and the reviewers for their commitment and support in completing this book.

We hope that our book will serve as a valuable handbook for molecular docking enthusiasts and promote future developments and breakthroughs in computational drug discovery and biomolecular modeling.

Črtomir Podlipnik
Faculty of Chemistry and Chemical Technology,
University of Ljubljana,
Ljubljana, Slovenia

Section 1

Methods

Chapter 1

Navigating the Computational Seas: A Comprehensive Overview of Molecular Docking Software in Drug Discovery

Mohd Mursal, Mohammad Ahmad, Sahil Hussain and Mohemmed Faraz Khan

Abstract

Molecular docking software is an indispensable tool in the dynamic landscape of computational biology and drug discovery. This software facilitates the simulation and prediction of interactions between small molecules and target proteins, offering a detailed glimpse into molecular binding phenomena. Employing sophisticated algorithms, these programs assess binding affinity, predict binding modes, and contribute crucial insights to streamline drug development processes. Examples of widely used molecular docking software include AutoDock, GOLD, and DOCK, each renowned for their capabilities in optimizing lead compounds, virtual screening, and elucidating the intricate details of protein-ligand interactions. Researchers leverage these tools to expedite the identification of potential drug candidates, ultimately bridging the gap between theoretical predictions and experimental advancements in the pursuit of novel therapeutics.

Keywords: molecular docking, drug design, bioinformatics, computational chemistry, binding affinity

1. Introduction

Amidst the binary symphony of code and the precision of algorithms, Molecular Docking Software emerges as a crucial instrument in the scientific toolkit, within the digital confines of computational biology. The landscape of molecular docking has evolved over the past thirty years, spurred by the demands of structural molecular biology and the pursuit of structure-based drug discovery. This evolution finds its roots in the remarkable expansion of computational capabilities, coupled with the increasing accessibility of databases housing small molecules and proteins. The synergy of these advancements has propelled the field forward, opening avenues for deeper exploration and understanding [1–3]. Automated molecular docking software sets out to unravel and foresee molecular recognition on two fronts: structurally, by identifying probable binding modes, and energetically, by forecasting binding affinity. This computational endeavor predominantly occurs in the dynamic interplay between a small molecule and a target macromolecule [4]. The objective is to decode the intricate language of molecular interactions, providing insights into the binding intricacies that govern biochemical processes. Frequently termed as ligand-protein docking, there is an increasing fascination with the exploration of protein–protein docking. The applications of molecular docking in drug discovery are diverse and encompass a spectrum of functions. These include conducting structure–activity studies, optimizing lead compounds, identifying potential leads through virtual screening, offering binding hypotheses to support predictions for mutagenesis studies, aiding x-ray crystallography by fitting substrates and inhibitors to electron density, facilitating chemical mechanism studies, and contributing to the design of combinatorial libraries. The versatility of molecular docking underscores its pivotal role across various stages of the drug development process [5, 6]. Simulations reveal biological disorder impacts (mutation effects, phosphorylation, protonation states) in various complexes: protein–protein, protein-ligand, protein-peptide, or protein-nucleic acid interactions [7]. In this section, our emphasis has been on molecular docking, the underlying algorithms, and the execution of simulations. Utilizing virtual screening grounded in molecular descriptors and physicochemical properties of both active and inactive ligands proves highly valuable in identifying hits and leads through library enrichment for screening [8]. This approach is also effective in streamlining and enhancing ligand libraries for molecular docking, with emerging evidence suggesting that ligand shape matching can be as effective as, if not more effective than, docking. Nonetheless, when molecular docking serves as the concluding phase in virtual screening, it contributes essential 3D (three-dimensional) structural hypotheses elucidating how a ligand engages with its target [9].

2. Molecular docking

Molecular docking uses a variety of SF (scoring functions), it is an in-silico technique that identifies the proper binding pose of a protein-ligand complex and assesses its strength to choose the optimal posture produced by each molecule to a rank order [10]. Docking techniques strive to seamlessly position a ligand within the binding site of a target protein. This involves the harmonization and optimization of variables such as hydrophobic, steric, and electrostatic complementarity, thereby providing an estimate of their binding free energy [11]. Molecular docking encapsulates three primary objectives: virtual screening, pose prediction, and the estimation of binding affinity [12]. A dependable docking technique should effectively distinguish between binding and non-binding sites, discerning their molecular interactions. Moreover, when handling extensive compound libraries, the method must aptly differentiate binding molecules from non-binding ones and accurately rank the binding molecules amongst the top compounds in the database [13]. The efficacy of virtual screening hinges on the quantity and precision of structural information available for the target protein and the ligand undergoing docking. The initial phase involves scrutinizing the target to identify the presence of pertinent binding sites [14]. This analysis is accomplished by studying established protein-ligand co-crystal structures or employing in-silico methods to pinpoint novel binding pockets [15]. Similar pockets are found using techniques such as the catalytic site AFT [16], SURFACE [17], POCKET SURFER, PATCH-SURFER, and Atlas. A "spherical probe," or a pre-computer depiction of an idealized ligand, is rolled over the grid surface of another in-silico approach called POCKET to identify the binding site of a target protein. Moreover, various small molecule databases, including ZINC, PubChem, ChemDB, and DrugBank, collectively host over a million deposited structures of potential ligands. Docking has become an essential virtual screening process, particularly when thousands or even millions of compounds need to be evaluated against one or more targets in a short amount of time from a database. Replicating such a search would be unfeasible in terms of time and money experimentally. In docking simulations, achieving a successful outcome relies on the pivotal features of speed and accuracy. The primary goal in developing a docking algorithm is to create a swift method capable of identifying novel lead compounds during virtual screening or accurately reproducing experimental conformations for validation with experimental data. Numerous docking programs include ZDOCK [18], M-ZDOCK [19], MS-DOCK [20], Surflex [21], MCDOCK [22], GOLD, AUTODOCK [23], and others.

Every docking program uses a different search technique, such as IC (Incremental Construction) [24], GA (Genetic Algorithm) [25], MC (Monte Carlo) [26], and others. As explained in more detail in the section that follows, each has a distinct parameter set and search strategy. The receptor and ligand input coordinates are used as a set of parameters by the docking algorithm, which searches for the best possible match between two or more molecules. These factors include the flexibility of receptor or ligand structures, as well as geometric complementarity that takes into consideration the charge and radius of atoms in the VDW (van der Waals). Interatomic interactions are also taken into account, including hydrophobic contacts and hydrogen bonds. The expected alignments (poses) of a ligand within the target's binding site are the result of docking applications. The posing usually results in several reasonable conformations. SFs that can assess binding free energy or intermolecular binding affinity are used to improve and rank these results. This optimization process culminates in the selection of the best orientation after the docking procedure.

3. Algorithms

Various molecular docking programs are constructed based on a range of algorithms, each offering unique advantages. This overview focuses on highlighting the most prominent algorithms and their associated benefits. Additionally, we provide insights into key docking programs developed around these algorithms. It is worth noting that these programs often leverage one or more specific search algorithms in their operations. To facilitate understanding, the categorization of these search algorithms, as depicted in **Figure 1**, serves as a structured framework for comprehending their functionalities and applications:

1. A **systematic or direct approach:** The following are the three subtypes of systematic methods:

    a. **Conformational search:** The structural property of the ligand is gradually adjusted in its torsional (dihedral), translational, and rotational degrees of freedom.

    b. **Fragmentation:** In this case, the fragments can be anchored individually, with the first fragment being docked initially and the remaining fragments being constructed outward in stages from that initial bound point, or they can be

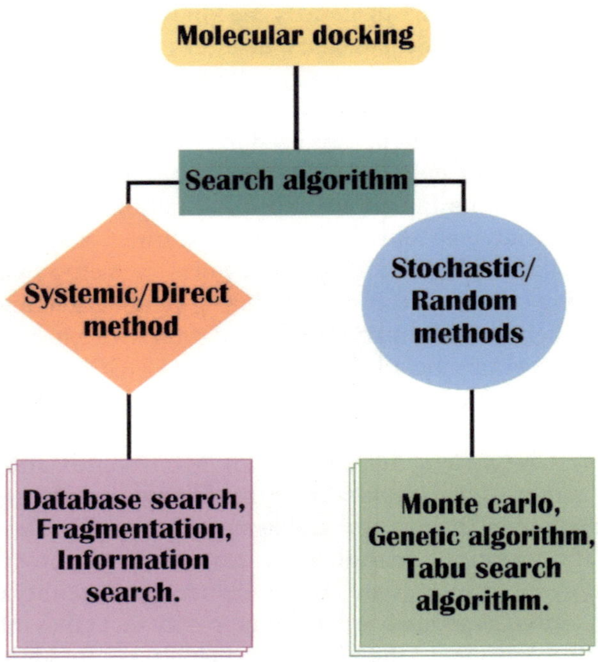

Figure 1.
Classification of the search algorithm in molecular docking.

docked together to create connections during the molecular docking process. It employs tools such as DOCK, LUDI, Flex XTM, etc.

c. **Database Search**: With this method, all the small molecules that are already registered in the database may be created in a large number of appropriate conformations, which can then be docked as hard bodies. One of the instruments it uses is FLOG.

2. **Stochastic methods or Random methods:** Three subcategories of stochastic approaches exist.

a. **Monte Carlo:** This method entails scoring the receptor binding site by randomly inserting ligands, after which a new configuration is produced. It uses instruments like ICM, MCDOCK, and so on.

b. **Genetic algorithm:** The "gene" describes the configuration and placement of the receptor concerning a population of postures, whilst the "fitness" determines the score. To create the following generation and repeat the agreement, carry out transformations, hybrids, etc. of the fittest. It employs applications like AutoDock, GOLD, and others.

c. **Tabu search:** It works by imposing restrictions that make it easier to study a new configuration by avoiding further examination of the ligands' conformational space's previously revealed regions. It makes use of MVD (Molegro Virtual Docker) TM, PRO LEADS, and other technologies [27].

3.1 Fast shape matching

Fast shape matching algorithms consider the geometric overlap between two molecules, employing various methods to generate alignments between ligand and receptor. This approach can identify potential binding sites of a protein through a macromolecular surface search. One prominent fast shape matching algorithm used in molecular docking is the Shape Complementarity principle, which evaluates the geometric fit between the ligand and the protein's binding site.

The application of fast shape matching algorithms significantly accelerates the initial screening process in molecular docking studies, enabling the rapid identification of potential ligands that fit well into the target protein's binding site. By efficiently filtering out poorly fitting ligands, these algorithms help prioritize promising candidates for further optimization and experimental validation. Additionally, they play a crucial role in virtual screening campaigns aimed at identifying lead compounds for drug discovery. Moreover, these algorithms establish plausible conformations of predicted binding sites [28]. For instance, the initial steps of the DOCK program involve delineating the regions of the binding site where a potential ligand could be situated. This is achieved by identifying sphere centres through a Matching algorithm. Notably, this approach is also utilized in many other widely recognized programs, including DOCK itself [29].

3.2 Monte Carlo algorithm

MC algorithms are stochastic computational methods used for sampling from probability distributions to solve various types of problems, including optimization and integration. In molecular docking, it is often employed to explore the conformational space of ligands and find energetically favorable binding poses with target proteins.

MC simulations were initially introduced as a minimization process in molecular dynamics applications, exemplified by implementations in GROMACS and GROMOS [30, 31]. Subsequently, they found adaptations for flexible docking algorithms like MCDOCK and ICM. Additionally, MC simulations have demonstrated efficacy in accurately determining relative binding constants for protein systems [32]. This method is great for studying different thermodynamic conditions in biomolecular systems, but it is not good for analyzing time-dependent processes like molecular dynamics simulations. But with flexible docking programs, the search algorithm tries out various orientations and rotations to dock the ligand into the receptor site, reducing the chance of getting stuck in local minima. Stochastic techniques are used to determine potential ligand orientations inside the binding site, generating semi-random number sequences in MC simulations. If needed, these sequence generators can recreate the sequence and make sets of independent integers.

The MC methods rely on the Metropolis algorithm, which furnishes an acceptance criterion in the evolution of the docking search. At each iteration of the algorithm, a random variation of the ligand's degrees of freedom is introduced [33]. The change is accepted when the energy score of the binding poses increases; otherwise, it is accepted based on the P (probability) expressed in the following equation.

$$P\,accept = \min\left(1, \exp\left(-kT\Delta E\right)\right)$$

Where, P accept is the probability of accepting a proposed conformation. ΔE is the change in energy associated with the proposed conformation. k is the Boltzmann constant. T is the temperature of the system [34].

3.3 Particle swarm optimization

Molecular docking employs swarm optimization techniques as a metaheuristic, serving as an effective search strategy for addressing the intricacies of the docking problem. Here, the docking problem is framed as an optimization task, where various parameters are associated with a well-defined SF. The primary objective is to ascertain the optimal conformation of a ligand within the receptor site, with the aim of minimizing energy interactions.

PSO (Particle Swarm Optimization) has emerged as a rapid and precise method uniquely suited to tackle the complexities inherent in molecular docking. This approach facilitates the exploration of the vast conformational space effectively, seeking the most favorable ligand-receptor interaction configurations. Numerous docking programs, such as SODOCK, are developed based on swarm principles, harnessing the collective intelligence of a swarm of particles to navigate the search space efficiently.

The PSO principle aligns seamlessly with the requirements of molecular docking, particularly emphasizing the minimization of the ligand's energy to optimize its binding affinity according to the specified SF. This synergy between PSO and molecular docking enhances the efficiency and accuracy of the docking process, making it a preferred choice in computational drug discovery and protein-ligand interaction studies.

3.4 Genetic algorithm

A GA is a search heuristic inspired by the principles of natural selection and genetics. It is commonly used to find approximate solutions to optimization and search problems. Molecular docking, on the other hand, is a computational technique used in structural biology and bioinformatics to predict how two or more molecules interact with each other.

The application of GA in molecular docking involves using the principles of natural selection to optimize the orientation and conformation of ligands within the binding site of a target protein. The goal is to find the best possible binding pose, which corresponds to the configuration where the ligand binds most tightly to the protein [35].

GA proves valuable in identifying the best parameters influencing the activity of the studied drug molecule and enhancing computational efficiency. Inspired by Darwin's theory of natural evolution, GA employs a genetic operator to combine two chromosomes (parents) and generate a new chromosome that may outperform its parents [36, 37]. This process incorporates multiple SF along with a set of parameters such as crossover rate and mutation rate. So there are many software applications that employ GA and one of them is GOLD (Genetic Optimization for Ligand Docking), which is a molecular docking software that employs a GA to explore the conformational space of ligands and predict their binding modes with a target protein.

4. Molecular docking tools

Molecular docking tools are computational programs designed to predict the binding mode and affinity of a ligand to its target protein. These tools employ various algorithms to explore the spatial and energetic complementarity between the ligand and the protein binding site. By simulating the docking process, they assist researchers in understanding the molecular interactions underlying biological processes and aid in drug discovery efforts. **Table 1** provides an overview of some well-known docking tools along with the algorithms they utilize. These tools play a crucial role in computational chemistry and biology, enabling the study of ligand-protein interactions in diverse biological systems.

4.1 GLIDE

Glide is a widely used molecular docking software developed by Schrödinger. It is renowned for its accuracy in predicting ligand-protein interactions and its efficient performance in virtual screening studies. It offers two precision modes: SP (Standard Precision) and XP (Extra Precision). These modes provide varying levels of accuracy and computational efficiency, allowing users to tailor their docking studies to their specific needs. Glide exploits advanced SF to evaluate and rank ligand binding poses. These SF consider various factors such as steric clashes, hydrogen bonding, and van der Waals interactions to assess the quality of ligand-protein interactions and can handle flexible ligands and protein receptor structures, allowing for more realistic modeling of molecular interactions. Glide is capable of efficiently screening large libraries of compounds to identify potential drug candidates or lead molecules [57]. Schrödinger provides a user-friendly graphical interface for setting up and running docking experiments with Glide, making it accessible to both experts and beginners in computational drug discovery.

4.2 AutoDock Vina

It is a powerful tool for predicting ligand-receptor interactions in drug discovery. Dr. Oleg Trott developed AutoDock Vina at The Scripps Research Institute. It estimates binding affinities between ligands and target proteins. The most recent stable version is 1.1.21. AutoDock Vina integrates GA and empirical scoring to efficiently explore ligand conformational space, considering steric clashes and electrostatic interactions for binding affinity estimation. Unlike rigid methods, it permits ligand partial flexibility, enhancing prediction accuracy. Moreover, it identifies potential binding sites in target proteins, aiding researchers in focusing on relevant regions. Recent advancements include QuickVina2 with enhanced search algorithms, Smina offering a user-friendly interface and customizable scoring terms, Vina-Carb optimized for carbohydrate docking, VinaXB addressing halogen bonding, and Vinardo incorporating an improved SF.

4.3 FlexX

FlexX is another molecular docking software, primarily used for ligand docking studies. Developed by BioSolveIT, FlexX offers several features tailored for docking small molecules into protein binding sites, particularly its ability to handle flexible ligands and protein structures. It can account for conformational changes in both the ligand and the receptor during docking simulations, which is crucial for accurately predicting ligand binding modes. FlexX employs SF to evaluate and rank ligand poses within the protein binding site. These SFs consider various factors, such as steric clashes, hydrogen bonding, and electrostatic interactions, to assess the quality

Docking tools	Algorithms	Benefits	Drawbacks	Reference
Glide	Hierarchical method	Its ability to efficiently explore ligand binding conformations based on geometric complementarity, offering a balance between speed and accuracy in molecular docking.	Its limited ability to account for dynamic flexibility in ligands and receptors, potentially leading to inaccuracies in predicting binding modes in highly flexible systems.	[38]
OEDocking v 3.0.0	Non-Stochastic method	Efficiently explore conformational space, ensuring comprehensive coverage and enhancing the likelihood of identifying precise ligand binding poses.	It may struggle to efficiently sample highly flexible ligands or receptor conformations, potentially leading to missed binding modes or inaccurate predictions in systems with significant conformational variability.	[39]
FlexX	LUDI	The fragment-based approach provides unbiased fragments to the active site of the receptor and utilizes the protein-ligand complex hydrogen bond.	It is difficult to find a template docked with pieces.	[40]
ZDOCK server, Patchdock	Fast Fourier transform algorithm	Docking rigid body proteins predicts how proteins and ligands interact, determining their energy. Worldwide, researchers explore diverse protein-ligand systems to understand their interactions better.	It offers a constrained perspective on frequencies in terms of signal processing.	[41, 42]
HOOK, FlexX	MCSS (Multiple Copy Simultaneous Search)	Functional ligand docked with several copies of the receptor binding site. Functional ligands should minimize their energy before docking.	MCCS is limited to use with rigid receptors.	[43]
MolDock, AutoDock, Lead finder	EP (Evolutionary programming)	Utilizing a heuristic search strategy, differential evolution is employed to anticipate cavities.	Only for flexible docking.	[44–46]
AUTODOCK VINA	Molecular dynamics	More accurately, the stability of the docked complex is represented.	Requiring a lot of time and computing resources and high barrier energy conformation that is challenging.	[47]
AutoDock4, MolDock	Simulated annealing	Analyzing conformation and flexibility Particular types of dynamic simulations When coupled with the Results from the MC approach are more accurate.	Time-consuming because of the simulation loop. It is necessary to combine MC and GA algorithms to provide precise outcomes.	[23]

Docking tools	Algorithms	Benefits	Drawbacks	Reference
eHiTS, DOCK 4.0, Hammerhead, FLOG, SLIDE	Incremental construction	Separate docking of the fragmented ligand at the receptor's active location.	If ligands have more rotatable bonds, then it fails.	[48, 49]
Pro leads, SFDock	Tabu search	A metaheuristic algorithm allows for the creation of new inquiries whilst preventing the consideration of previously evaluated solutions.	It is employed in the mean square error computation.	[50, 51]
ClustMPSO	PSO (Particle swarm optimization)	A quick heuristic approach is more reliable.	The uncertainty of convergence is an issue in PSO.	[52]
GOLD 3.1, PSI-DOCK, rDOCK	GA	Utilize the advantages of parallelization for both rigid and flexible ligand applications.	For time-dependent methods, such as those used in molecular dynamics simulations, it is not appropriate. There is a problem with the GA's convergence uncertainty.	[53]
MCDOCK, RiboDock, Glide, ROSETTALIGAND	MC	Precise and exact execution	For time-dependent methods, such as those used in molecular dynamics simulations, it is not appropriate.	[54]
Ph4DOCK, GM-DockZn	Matching algorithm	Advantage of speed	Prior knowledge of the specifics of the receptor locations is required. The precision is impacted by the molecular flexibility lacking.	[55, 56]

Table 1.
Docking tools, algorithms, benefits, and drawbacks.

of ligand-protein interactions. FlexX is designed to rapidly explore the conformational space of ligands and protein receptors, making it suitable for high-throughput virtual screening studies. FlexX provides a user-friendly interface for setting up docking experiments and analyzing results. It offers visualization tools to help users understand and interpret docking outcomes. FlexX allows users to customize various parameters of the docking process, such as search algorithms, SF, and constraints, to tailor the docking studies to their specific research needs [58].

4.4 GOLD

GOLD (Genetic Optimization for Ligand Docking) is a molecular docking software developed by the CCDC (Cambridge Crystallographic Data Centre). It is widely used in the pharmaceutical industry and academic research for virtual screening, lead optimization, and structure-based drug design. GOLD employs a GA to explore ligand conformational space and optimize ligand binding poses within protein binding sites. The GA mimics natural selection processes to iteratively improve ligand binding affinity and geometry. In molecular docking, GOLD uses SF to rank ligand poses based on their predicted binding affinity and compatibility with the protein's binding site. These SFs take into account factors like steric clashes, hydrogen bonding, and hydrophobic interactions. It can handle flexible ligands and protein receptor structures, allowing for accurate modeling of conformational changes during ligand binding. It can also incorporate flexibility into specific regions of the protein binding site to improve docking accuracy. The latest iteration of GOLD Suite 5.2 encompasses three key components: Gold 5.2 facilitates protein-ligand docking, Hermes 1.6 enables comprehensive protein visualization, and Gold Mine 1.5 offers tools for the analysis of docking results, including grading.

4.5 LeDOCK

LeDock is a molecular docking software developed by the Laboratory of Structural Bioinformatics at the University of Paris Diderot and the Molecular Modeling Group (CNRS, France). LeDock employs a combination of geometric matching and optimization algorithms for molecular docking. It utilizes geometric complementarity to match ligand conformations to the protein binding site and optimizes ligand poses to improve binding affinity. Evaluation and ranking of ligand poses are carried out by LeDock using SF, considering factors such as steric clashes, hydrogen bonding, and electrostatic interactions. It accommodates flexible ligands and protein receptor structures, allowing for accurate modeling of conformational changes during ligand binding. Robust performance has been demonstrated in virtual screening and lead optimization studies across various protein targets. It has been validated against experimental data and benchmarked against other docking software. LeDock is open-source software, that grants users access to and modification of the source code for customization or integration into larger computational pipelines [59].

5. Scoring function

An SF is a computational model used in molecular docking to evaluate and rank the binding interactions between a ligand and a protein. The SF takes into account various molecular characteristics, such as the steric fit, electrostatic interactions, hydrogen bonding, and other factors that contribute to the stability of the ligand-protein complex. The primary goal of an SF is to predict and prioritize the binding affinity or strength of different ligands to a given target, helping researchers identify potential drug candidates or understand the molecular basis of biological processes. Different molecular docking programs may employ distinct SFs, and their accuracy is crucial for the reliability of the virtual screening results [60].

5.1 Physics-based scoring functions

A physics-based SF is a sophisticated framework that analyses molecular interactions with a high degree of accuracy. It comprises three integral components, each contributing to a comprehensive understanding of ligand-protein complexes: force field, solvent model, and quantum mechanics. The force field aspect employs a mathematical model to calculate the potential energy of a molecular system. It includes parameters for bond stretching, angle bending, and dihedral angle rotation, providing a classical representation of molecular mechanics. By capturing the geometric aspects and internal energy of the ligand-protein complex, the force field

DOI: http://dx.doi.org/10.5772/intechopen.1004802

enhances the overall accuracy of the SF. Solvent models are incorporated to account for the impact of the surrounding environment, typically water, on the molecular system. These models consider the effects of solvation on the complex's energetics. Integrating solvent models into the SF enhances the realism of the molecular environment, taking into account the impact of water molecules on the stability of complexes. This incorporation acknowledges the role of solvent effects in influencing the interactions within the molecular system. Quantum mechanics methods, such as DFT (density functional theory) or ab initio calculations, are employed to compute the electronic structure of the ligand-protein complex. The quantum mechanics component provides a highly accurate representation of electronic interactions, including bond formation, charge distribution, and electronic energy contributions, contributing to a detailed assessment of molecular energetics [61].

5.2 Empirical scoring functions

Empirical SF evaluates a complex's binding affinity by combining important energetic components associated with protein-ligand binding, such as steric clashes, hydrophobic effects, and hydrogen bonds, amongst others. Usually, linear regression analysis is employed to enhance the weights of these energy components using a training set with known binding affinities. Empirical SFs encompass two primary research directions. The first direction involves the utilization of extensive and high-quality training datasets to enhance the optimization of protein-ligand structures. The second direction focuses on selecting suitable energy terms through stepwise variables and systematic considerations for the target protein. Whilst empirical SFs break down protein-ligand binding affinities into distinct energy terms, akin to physics-based SFs, they typically adopt a flexible and intuitive functional form rather than relying on the well-established models employed by physics-based SFs. Due to their straightforward energy terms, these SFs excel in predicting binding affinity, ligand pose, and virtual screening with minimal computational expense. However, they are less adept at elucidating the correlation between binding affinity and crystal structures, and they often face challenges related to double-counting issues [62].

5.3 Knowledge-based scoring functions

Knowledge-based SFs are computational models used in molecular docking and structure-based drug design to predict the binding affinity of a ligand to a target protein. Unlike physics-based SFs that rely on theoretical models of molecular interactions, knowledge-based SFs derive their information from experimental data and existing knowledge about protein-ligand interactions. The key idea behind knowledge-based SFs is to utilize statistical analysis of experimental structures to identify patterns and preferences in protein-ligand interactions. These functions are trained on databases of experimentally determined protein-ligand complexes, such as the PDB (Protein Data Bank). The statistical analysis involves extracting information about the frequency and distribution of various molecular interactions observed in these complexes [63].

This is because SFs share several common features. Firstly, they predominantly focus on pairwise interactions between specific atoms or atom types in both the ligand and the protein. This allows the SF to pinpoint and evaluate the individual contributions of these interactions. Secondly, these SFs rely on statistical potentials to assess the likelihood of observing a particular interaction at a given distance or angle. By drawing on statistical analyses of experimental structures, they derive insights into the preferences and tendencies of real protein-ligand interactions. Additionally, knowledge-based SF takes into account the distributions of distances and angles between interacting atoms observed in experimental structures. This consideration of spatial arrangements contributes to a more accurate assessment of the likelihood and strength of various interactions. Furthermore, the SF encompasses energy terms that encompass a range of molecular forces and interactions. These terms may include contributions from van der Waals interactions, hydrogen bonding, electrostatic interactions, and other relevant energy components, providing a comprehensive evaluation of the molecular binding. Lastly, some knowledge-based SF incorporates a reference state, which represents the expected behavior of non-interacting atoms. This reference state serves as a baseline for evaluating the significance and favourability of observed interactions within the context of the specific molecular environment [64].

5.4 Machine learning-based scoring functions

Machine learning-based SF is a computational model that utilizes machine learning techniques to predict the binding affinity or likelihood of interaction between a protein and a ligand in molecular docking studies. Molecular docking is a computational method used in drug discovery to analyze and predict the preferred orientation and conformation of a ligand when binding to a target protein. These models are trained on datasets containing information about known protein-ligand complexes as shown in **Figure 2**, including their experimentally measured binding affinities. The machine learning algorithm learns patterns and relationships between various features or descriptors associated with these complexes. These features could include structural, energetic, or physicochemical properties of the protein and ligand. Once trained, the SF can be applied to predict the binding affinity of new

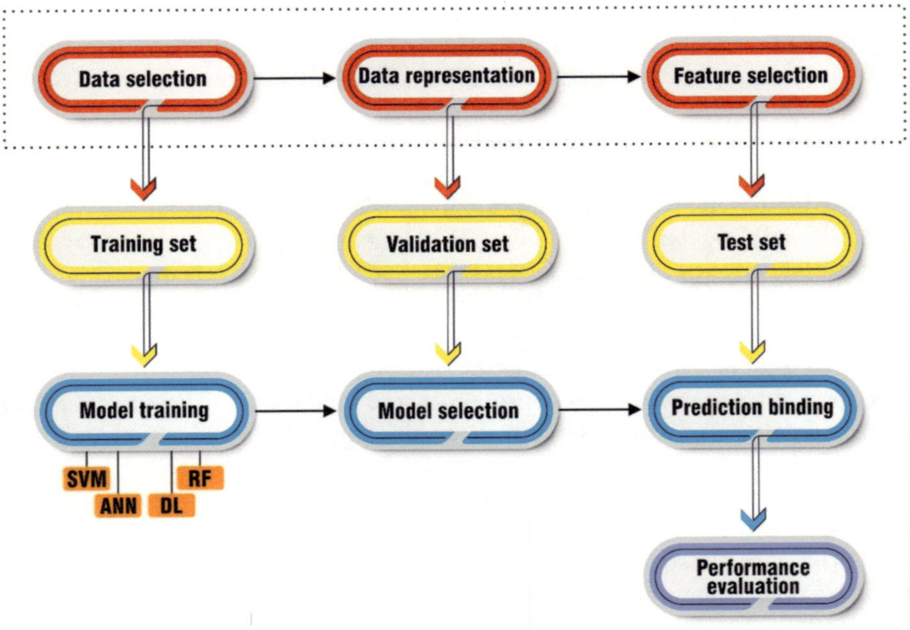

Figure 2.
Training process for a machine learning-based SF.

protein-ligand interactions that were not part of the training dataset. The goal is to provide accurate and efficient predictions, aiding in the identification of potential drug candidates or understanding the strength of interactions between biological molecules. Common machine learning algorithms used for developing SF include linear regression, support vector machines, random forests, and more recently, deep learning approaches such as neural networks. Machine learning-based SFs are valuable tools in drug discovery and computational biology, helping researchers prioritize and analyze potential drug candidates more effectively [65].

6. Steps in molecular docking

Outlined below are the key stages in molecular docking, a pivotal process in structure-based drug design. These steps systematically guide the preparation, execution, and analysis of ligand-protein interactions, providing a roadmap for elucidating molecular binding mechanisms for a visual representation of these steps, refer to **Figure 3**, which illustrates the general workflow of molecular docking:

a. **Preparation of protein and ligand structures**: Retrieve or generate the 3D structures of the protein receptor and the ligand molecule to be docked. Prepare the protein structure by removing water molecules, heteroatoms, and co-crystallized ligands. Also, check and correct any structural irregularities. Prepare the ligand structure by adding hydrogen atoms, assigning proper charges, and optimizing its geometry [66, 67].

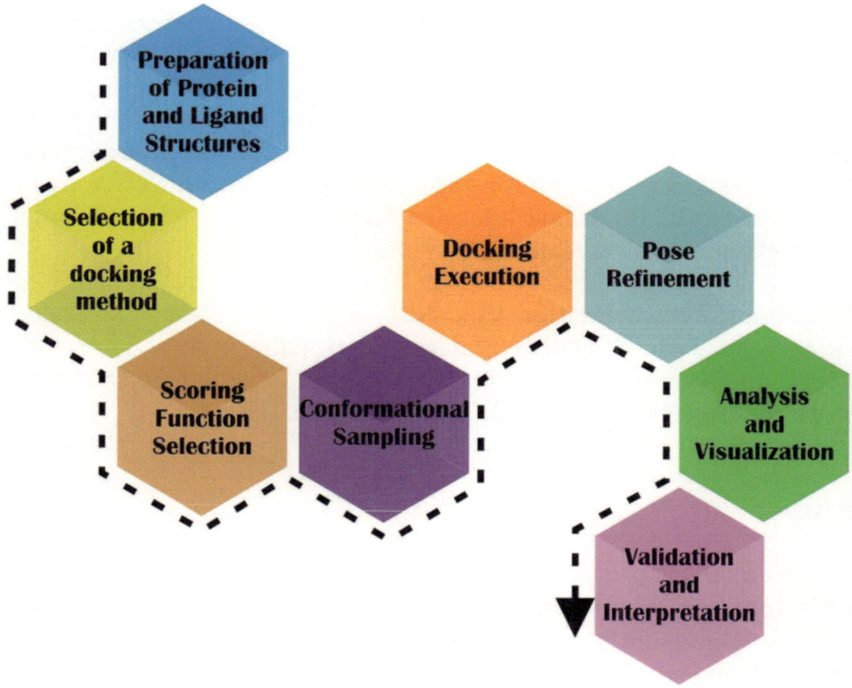

Figure 3.
General steps involved in molecular docking.

b. **Selection of a docking method:** It hinges on the researcher's requirements. For instance, if docking numerous molecules at a protein's binding site under specific conditions like pH, water presence, and solubility is the aim, flexible docking programs might be favored. However, when scanning thousands of compounds from databases, flexible docking methods could be impractical unless there is ample processing power and a speedy computer. Hence, users can opt for various docking methods based on their computer's capabilities and the properties of the target [68].

c. **Scoring function selection**: Choose an appropriate SF or scoring scheme to evaluate the fitness of ligand poses within the binding site. SF assesses the binding affinity based on various parameters such as van der Waals interactions, electrostatics, hydrogen bonding, and solvation effects.

d. **Conformational sampling**: Perform a systematic search of the ligand conformational space to explore different possible binding poses within the defined search space. Use algorithms such as GA, MC methods, or molecular dynamics simulations to generate and optimize ligand conformations [69].

e. **Docking execution**: Apply a docking algorithm to position the ligand within the binding site based on the calculated energy scores. Commonly used docking algorithms include AutoDock, DOCK, Vina, Glide, and GOLD, amongst others [70].

f. **Pose refinement**: Refine the initial ligand poses obtained from docking algorithms to improve accuracy and reliability. Use methods like molecular dynamics simulations, energy minimization, or rescoring techniques to refine the ligand-protein interactions [71].

g. **Analysis and visualization**: Analyze the docked poses to identify potential binding modes and interactions between the ligand and the protein receptor. Visualize the docked complexes using molecular visualization software to interpret the results and make informed decisions regarding ligand design or further experimental validation [72].

h. **Validation and interpretation**: Validate the docking results by comparing them to experimental data, if available. Interpret the binding poses and interactions to understand the molecular mechanisms underlying ligand binding and to guide further experimental studies or drug design efforts [73].

7. Conclusion

In summary, this chapter provides a thorough exploration of molecular docking software in drug discovery within the realm of computational biology and drug design. Molecular docking emerges as a vital tool, predicting interactions between small molecules and target proteins. Over the last three decades, its evolution has been propelled by computational advancements and increased database accessibility. Our discussion highlights molecular docking's pivotal role in drug development, from structure–activity studies to lead optimization and virtual screening. These software tools, including AutoDock, GOLD, and DOCK, play crucial roles in understanding molecular interactions. We explore algorithms and search strategies, ranging from systematic approaches such as conformational search to stochastic methods like MC and GA. Shape matching algorithms, MC simulations, particle swarm optimization, and GA are investigated, providing insights into their applications and benefits. SF, spanning physics-based, empirical, knowledge-based, and machine learning-based approaches are highlighted for their roles in evaluating binding interactions. This chapter serves as a guide for researchers, empowering them to navigate the complexities of molecular docking for innovative drug discoveries in the ever-evolving field of computational biology.

Author details

Mohd Mursal, Mohammad Ahmad, Sahil Hussain and Mohemmed Faraz Khan*
Faculty of Pharmacy, Integral University, Lucknow, Uttar Pradesh, India

*Address all correspondence to: faraz91khan@gmail.com

References

[1] Hu L, Benson ML, Smith RD, Lerner MG, Carlson HA. Binding MOAD (Mother of All Databases). Proteins: Structure, Function and Genetics. 2005;**60**(3):333-340. DOI: 10.1002/PROT.20512

[2] Ahmed A, Mam B, Sowdhamini R. DEELIG: A deep learning approach to predict protein-ligand binding affinity. Bioinformatics and Biology Insights. 2021;**15**:2-5. DOI: 10.1177/11779322211030364

[3] Roche O, Kiyama R, Brooks CL. Ligand-protein database: Linking protein-ligand complex structures to binding data. Journal of Medicinal Chemistry. 2001;**44**(22):3592-3598. DOI: 10.1021/JM000467K/ASSET/IMAGES/LARGE/JM000467KN00001.JPEG

[4] Cournia Z, Allen B, Sherman W. Relative binding free energy calculations in drug discovery: Recent advances and practical considerations. Journal of Chemical Information and Modeling. 2017;**57**(12):2911-2937. DOI: 10.1021/ACS.JCIM.7B00564

[5] Maruca A et al. Computer-based techniques for lead identification and optimization i: Basics. Physical Sciences Reviews. Jun 2019;**4**(6):4-7. DOI: 10.1515/PSR-2018-0113

[6] Kontoyianni M. Docking and virtual screening in drug discovery. Methods in Molecular Biology. 2017;**1647**:255-266. DOI: 10.1007/978-1-4939-7201-2_18

[7] Hollingsworth SA, Dror RO. Molecular dynamics simulation for all. Neuron. 2018;**99**(6):1129-1143. DOI: 10.1016/j.neuron.2018.08.011

[8] Pozzan A. Molecular descriptors and methods for ligand based virtual high throughput screening in drug discovery. Current Pharmaceutical Design. 2006;**12**(17):2099-2110. DOI: 10.2174/138161206777585247

[9] Hawkins PCD, Skillman AG, Nicholls A. Comparison of shape-matching and docking as virtual screening tools. Journal of Medicinal Chemistry. 2007;**50**(1):74-82. DOI: 10.1021/JM0603365

[10] Kuntz ID, Blaney JM, Oatley SJ, Langridge R, Ferrin TE. A geometric approach to macromolecule-ligand interactions. Journal of Molecular Biology. 1982;**161**(2):269-288. DOI: 10.1016/0022-2836(82)90153-X

[11] Younus S, Vinod Chandra SS, Nair ASS. Docking and dynamic simulation study of Crizotinib and Temozolomide drug with Glioblastoma and NSCLC target to identify better efficacy of the drug. Future Journal of Pharmaceutical Sciences. Dec 2021;**7**(1):6-8. DOI: 10.1186/S43094-021-00323-2

[12] Jain AN, Nicholls A. Recommendations for evaluation of computational methods. Journal of Computer-Aided Molecular Design. 2008;**22**(3-4):133-139. DOI: 10.1007/S10822-008-9196-5

[13] Kolb P, Irwin J. Docking screens: Right for the right reasons? Current Topics in Medicinal Chemistry. 2009;**9**(9):755-770. DOI: 10.2174/156802609789207091

[14] Kuan J, Radaeva M, Avenido A, Cherkasov A, Gentile F. Keeping pace with the explosive growth of chemical

libraries with structure-based virtual screening. Wiley Interdisciplinary Review in Computer Molecular Science. Nov 2023;**13**(6):4-10. DOI: 10.1002/WCMS.1678

[15] Freidel MR, Armen RS. Mapping major SARS-CoV-2 drug targets and assessment of druggability using computational fragment screening: Identification of an allosteric small-molecule binding site on the Nsp13 helicase. PLoS One. Feb 2021;**16**(2):2-5. DOI: 10.1371/JOURNAL.PONE.0246181

[16] Arakaki AK, Zhang Y, Skolnick J. Large-scale assessment of the utility of low-resolution protein structures for biochemical function assignment. Bioinformatics. 2004;**20**(7):1087-1096. DOI: 10.1093/BIOINFORMATICS/BTH044

[17] Ferrè F, Ausiello G, Zanzoni A, Helmer-Citterich M. SURFACE: A database of protein surface regions for functional annotation. Nucleic Acids Research. 2004;**32**:2-3. DOI: 10.1093/nar/gkh054

[18] Chen R, Li L, Weng Z. ZDOCK: An initial-stage protein-docking algorithm. Proteins: Structure, Function and Genetics. 2003;**52**(1):80-87. DOI: 10.1002/PROT.10389

[19] Pierce B, Tong W, Weng Z. M-ZDOCK: A grid-based approach for Cn symmetric multimer docking. Bioinformatics. 2005;**21**(8):1472-1478. DOI: 10.1093/bioinformatics/bti229

[20] Sauton N, Lagorce D, Villoutreix BO, Miteva MA. MS-DOCK: Accurate multiple conformation generator and rigid docking protocol for multi-step virtual ligand screening. BMC Bioinformatics. 2008;**9**(1):1-12. DOI: 10.1186/1471-2105-9-184/FIGURES/3

[21] Jain AN. Surflex: Fully automatic flexible molecular docking using a molecular similarity-based search engine. Journal of Medicinal Chemistry. 2003;**46**(4):499-511. DOI: 10.1021/JM020406H/ASSET/IMAGES/MEDIUM/JM020406HN00001.GIF

[22] Temml V, Kutil Z. Structure-based molecular modeling in SAR analysis and lead optimization. Computational and Structural Biotechnology Journal. 2021;**19**:1431-1444. DOI: 10.1016/j.csbj.2021.02.018

[23] Bitencourt-Ferreira G, Pintro VO, de Azevedo WF. Docking with AutoDock4. Methods in Molecular Biology. 2019;**2053**:125-148. DOI: 10.1007/978-1-4939-9752-7_9

[24] Fu Y, Wu X, Chen Z, Sun J, Zhao J, Xu W. A new approach for flexible molecular docking based on swarm intelligence. Mathematical Problems in Engineering. 2015;**34**:2-6. DOI: 10.1155/2015/540186

[25] Gardiner EJ, Willett P, Artymiuk PJ. Protein docking using a genetic algorithm. Proteins: Structure, Function and Genetics. 2001;**44**(1):44-56. DOI: 10.1002/PROT.1070

[26] Liu M, Wang S. MCDOCK: A Monte Carlo simulation approach to the molecular docking problem. Journal of Computer-Aided Molecular Design. 1999;**13**(5):435-451. DOI: 10.1023/A:1008005918983

[27] Kodchakorn K, Poovorawan Y, Suwannakarn K, Kongtawelert P. Molecular modelling investigation for drugs and nutraceuticals against protease of SARS-CoV-2. Journal of Molecular Graphics & Modelling. Dec 2020;**101**:2-6. DOI: 10.1016/j.jmgm.2020.107717

[28] Shoichet BK, Kuntz ID, Bodian DL. Molecular docking

using shape descriptors. Journal of Computational Chemistry. 1992;**13**(3):380-397. DOI: 10.1002/JCC.540130311

[29] Yamagishi M, Martins N, Neshich G, et al. A fast surface-matching procedure for protein–ligand docking. Springer. 2006;**12**(6):965-972. DOI: 10.1007/s00894-006-0109-z

[30] Van Der Spoel D, Lindahl E, Hess B, Groenhof G, Mark AE, Berendsen HJC. GROMACS: Fast, flexible, and free. Journal of Computational Chemistry. 2005;**26**(16):1701-1718. DOI: 10.1002/JCC.20291

[31] Meyer B, Möller H. Conformation of glycopeptides and glycoproteins. Topics in Current Chemistry. 2006;**267**:187-251. DOI: 10.1007/128_2006_078

[32] Jorgensen WL. Efficient drug lead discovery and optimization. Accounts of Chemical Research. Jun 2009;**42**(6):724. DOI: 10.1021/AR800236T

[33] Hart T et al. A multiple-start Monte Carlo docking method. Wiley Online Library. 1992. [Online]. Available from: https://onlinelibrary.wiley.com/doi/abs/10.1002/prot.340130304 [Accessed: January 06, 2024]

[34] Read R, Hart T, et al. Monte Carlo algorithms for docking to proteins. Taylor & Francis; [Online] 1994. pp. 3-4. Available from: https://www.tandfonline.com/doi/abs/10.1080/10610279508032529 [Accessed: January 06, 2024]

[35] Tüzün B, Saripinar E. Molecular docking and 4D-QSAR model of methanone derivatives by electron conformational-genetic algorithm method. Journal of the Iranian Chemical Society. 2020;**17**(5):985-1000. DOI: 10.1007/S13738-019-01835-8

[36] Katoch S, Chauhan SS, Kumar V. A review on genetic algorithm: Past, present, and future. Multimedia Tools and Applications. 2021;**80**(5):8091-8126. DOI: 10.1007/S11042-020-10139-6

[37] Prasad KK, Ghosh A. A genetic algorithm approach to optimal Asset allocation of defined contribution pension funds: Evidence from India's national pension system. Compensation and Benefits Review. 2023;**35**:2-3. DOI: 10.1177/08863687231195498

[38] Chen T, Shu X, Zhou H, Beckford FA, Misir M. Algorithm selection for protein–ligand docking: Strategies and analysis on ACE. Scientific Reports 2023. 2023;**13**(1):1-15. DOI: 10.1038/s41598-023-35132-5

[39] McGann M. FRED and HYBRID docking performance on standardized datasets. Journal of Computer-Aided Molecular Design. 2012;**26**(8):897-906. DOI: 10.1007/s10822-012-9584-8

[40] Pagadala NS, Syed K, Tuszynski J. Software for molecular docking: A review. Biophysical Reviews. 2017;**9**(2):91. DOI: 10.1007/S12551-016-0247-1

[41] Agrawal P, Singh H, Srivastava HK, Singh S, Kishore G, Raghava GPS. Benchmarking of different molecular docking methods for protein-peptide docking. BMC Bioinformatics. 2019;**19**: 7-8. DOI: 10.1186/S12859-018-2449-Y

[42] Schneidman-Duhovny D, Inbar Y, Nussinov R, Wolfson HJ. PatchDock and SymmDock: Servers for rigid and symmetric docking. Nucleic Acids Research. 2005;**33**(Web Server issue): W363. DOI: 10.1093/NAR/GKI481

[43] Eisen MB, Wiley DC, Karplus M, Hubbard RE. HOOK: A program for finding novel molecular architectures

that satisfy the chemical and steric requirements of a macromolecule binding site. Proteins. 1994;**19**(3): 199-221. DOI: 10.1002/PROT.340190305

[44] Park H, Lee J, Lee S. Critical assessment of the automated AutoDock as a new docking tool for virtual screening. Proteins: Structure, Function and Genetics. 2006;**65**(3):549-554. DOI: 10.1002/PROT.21183

[45] Thomsen R, Christensen MH. MolDock: A new technique for high-accuracy molecular docking. Journal of Medicinal Chemistry. 2006;**49**(11):3315-3321. DOI: 10.1021/JM051197E/SUPPL_FILE/JM051197ESI20060314_081922.PDF

[46] Gumede NJ. Pathfinder-driven chemical space exploration and multiparameter optimization in tandem with Glide/IFD and QSAR-based active learning approach to prioritize design ideas for FEP+ calculations of SARS-CoV-2 PLpro inhibitors. Molecules. 2022;**27**(23):8569. DOI: 10.3390/MOLECULES27238569/S1

[47] Eberhardt J, Santos-Martins D, Tillack AF, Forli S. AutoDock Vina 1.2.0: New docking methods, expanded force field, and Python bindings. Journal of Chemical Information and Modeling. 2021;**61**(8):3891-3898. DOI: 10.1021/ACS.JCIM.1C00203

[48] Zsoldos Z, Reid D, Simon A, Sadjad SB, Johnson AP. eHiTS: A new fast, exhaustive flexible ligand docking system. Journal of Molecular Graphics & Modelling. 2007;**26**(1):198-212. DOI: 10.1016/J.JMGM.2006.06.002

[49] Ewing TJA, Makino S, Skillman AG, Kuntz ID. DOCK 4.0: Search strategies for automated molecular docking of flexible molecule databases. Journal of Computer-Aided Molecular Design. 2001;**15**(5):411-428. DOI: 10.1023/A:1011115820450

[50] Fogel GB, Cheung M, Pittman E, Hecht D. Modeling the inhibition of quadruple mutant plasmodium falciparum dihydrofolate reductase by pyrimethamine derivatives. Journal of Computer-Aided Molecular Design. 2008;**22**(1):29-38. DOI: 10.1007/S10822-007-9152-9

[51] Hou T, Wang J, Chen L, Xu X. Automated docking of peptides and proteins by using a genetic algorithm combined with a tabu search. Protein Engineering, Design and Selection. 1999;**12**(8):639-648. DOI: 10.1093/PROTEIN/12.8.639

[52] Janson S, Merkle D, Middendorf M. Molecular docking with multi-objective particle swarm optimization. Applied Soft Computing. 2008;**8**(1):666-675. DOI: 10.1016/J.ASOC.2007.05.005

[53] Kellenberger E, Foata N, et al. Ranking targets in structure-based virtual screening of three-dimensional protein libraries: Methods and problems. ACS Publications. 2008;**48**(5):1014-1025. DOI: 10.1021/ci800023x

[54] Singh S, Bani Baker Q, Singh DB. Molecular docking and molecular dynamics simulation. Bioinformatics: Methods and Applications. Jan 2021;**18**:291-304. DOI: 10.1016/B978-0-323-89775-4.00014-6

[55] Goto J, Kataoka R, Hirayama N. Ph4Dock: Pharmacophore-based protein - ligand docking. Journal of Medicinal Chemistry. 2004;**47**(27):6804-6811. DOI: 10.1021/JM0493818/ASSET/IMAGES/MEDIUM/JM0493818N00001.GIF

[56] Wang et al. GM-DockZn: A geometry matching-based docking algorithm for

zinc proteins. Europe PMC. [Online]. Available from: https://europepmc.org/article/med/32369579 [Accessed: January 09, 2024]

[57] Friesner RA et al. Glide: A new approach for rapid, accurate docking and scoring. 1. Method and assessment of docking accuracy. Journal of Medicinal Chemistry. 2004;**47**(7):1739-1749. DOI: 10.1021/JM0306430/SUPPL_FILE/JM0306430_S.PDF

[58] Rarey M, Kramer B, Lengauer T, Klebe G. A fast flexible docking method using an incremental construction algorithm. Journal of Molecular Biology. 1996;**261**(3):470-489. DOI: 10.1006/jmbi.1996.0477

[59] Velavan S, Karnan R, NK-A J of Innovative, and undefined 2020, A comparative study on In silico software's in statistical relation to molecular docking scores. asianjir.com. 1997;**3**:1-5. Available from: https://asianjir.com/images/issues/BIOINFORMATICS.pdf [Accessed: January 27, 2024]

[60] Li J, Fu A, Zhang L. An overview of scoring functions used for protein–ligand interactions in molecular docking. Interdisciplinary Sciences. 2019;**11**(2):320-328. DOI: 10.1007/S12539-019-00327-W

[61] Meli R, Morris GM, Biggin PC. Scoring functions for protein-ligand binding affinity prediction using Structure-based deep learning: A review. Frontiers in Bioinformatics. 2022;**2**:2-3. DOI: 10.3389/FBINF.2022.885983

[62] Guedes IA, Pereira FSS, Dardenne LE. Empirical scoring functions for structure-based virtual screening: Applications, critical aspects, and challenges. Frontiers in Pharmacology. Sep 2018;**9**(sep):9-11. DOI: 10.3389/FPHAR.2018.01089

[63] Dittrich J, Schmidt D, et al. Converging a knowledge-based scoring function. ACS. 2018;**59**(1):509-521. DOI: 10.1021/acs.jcim.8b00582

[64] Pfeffer P et al. DrugScore RNA knowledge-based scoring function to predict RNA–ligand interactions. ACS. 2007;**47**(5):1868-1876. DOI: 10.1021/ci700134p

[65] Fujimoto KJ, Minami S, Yanai T. Machine-learning- and knowledge-based scoring functions incorporating ligand and protein fingerprints. ACS Omega. 2022;**7**(22):19030-19039. DOI: 10.1021/ACSOMEGA.2C02822

[66] Madhavi Sastry G, Adzhigirey M, Day T, Annabhimoju R, Sherman W. Protein and ligand preparation: Parameters, protocols, and influence on virtual screening enrichments. Journal of Computer-Aided Molecular Design. 2013;**27**(3):221-234. DOI: 10.1007/S10822-013-9644-8

[67] Aja PM et al. Prospect into therapeutic potentials of Moringa oleifera phytocompounds against cancer upsurge: de novo synthesis of test compounds, molecular docking, and ADMET studies. Bulletin of the National Research Centre. Dec 2021;**45**(1). DOI: 10.1186/S42269-021-00554-6

[68] Kitchen DB, Decornez H, Furr JR, Bajorath J. Docking and scoring in virtual screening for drug discovery: Methods and applications. Nature Reviews. Drug Discovery. 2004;**3**(11):935-949. DOI: 10.1038/nrd1549

[69] Elokely KM, Doerksen RJ. Docking challenge: Protein sampling and molecular docking performance. Journal of Chemical Information and Modeling. 2013;**53**(8):1934-1945. DOI: 10.1021/CI400040D

[70] Tiwari R, Prakash K, Pradesh U, Professor A, Tiwari S. Unrevealing the complex interplay: Molecular docking: A Comprehensive review on current scenario, upcoming difficulties. *ijcr.info*. 2024. 10.22159/ijcr.2024v8i1.226

[71] Yamashita T, Sakano T, Iqbal Mahamood M, Fujitani H. Molecular dynamics analysis to evaluate docking pose prediction. jstage.jst.go.jp. 2016;**13**:181-194. DOI: 10.2142/biophysico.13.0_181

[72] Agu PC et al. Molecular docking as a tool for the discovery of molecular targets of nutraceuticals in diseases management. Scientific Reports. 2023;**13**(1):13398. DOI: 10.1038/S41598-023-40160-2

[73] Klepeis JL, Lindorff-Larsen K, Dror RO, Shaw DE. Long-timescale molecular dynamics simulations of protein structure and function. Current Opinion in Structural Biology. 2009;**19**(2):120-127. DOI: 10.1016/j.sbi.2009.03.004

Chapter 2

In Silico Docking: Protocols for Computational Exploration of Molecular Interactions

Neha Mathur, Siva Sai Chandragiri, Sarita, Shristhi Shandily, Krupa Mukeshbhai Santoki, Nandini Navinchandra Vadhavana, Sejal Shah and Muktesh Chandra

Abstract

In computational chemistry and drug development, in silico docking has become an indispensable tool for investigating the molecular interactions between ligands and receptors. The procedures and approaches used in in-silico docking to decipher the complex dynamics of molecular binding processes are highlighted in this chapter. The first section of the chapter explains the basic ideas behind molecular docking, focusing on the function of scoring functions and algorithms in ligand-receptor interaction prediction. The benefits and drawbacks of several docking techniques—such as flexible docking, rigid-body docking, and other docking methods—are thoroughly covered. In addition, the challenges associated with conformational flexibility, solvent effects, and ligand desolvation that arise during in-silico docking are explored. Molecular dynamics simulations and ensemble docking techniques are investigated as ways to improve the precision and dependability of docking predictions. Furthermore, using in silico docking in virtual screening, structure-based drug design, and drug discovery highlights how important it is to speed up the drug development process and reduce experimental expenses. A thorough review of in silico docking techniques is given in this chapter, along with an examination of its methodological complexities, theoretical underpinnings, and real-world uses in drug discovery and computational chemistry.

Keywords: molecular docking, protein structure, virtual screening, QSAR, binding energy

1. Introduction

The completion of the human genome project has led to an increasing number of new therapeutic targets for drug discovery. Simultaneously, advancements in high-throughput protein purification, crystallography, and nuclear magnetic resonance spectroscopy techniques have provided detailed insights into proteins and protein-ligand complexes. These breakthroughs have enabled computational strategies to play a pivotal role in various aspects of drug discovery today [1–5], including virtual screening (VS) techniques [6] for hit identification and methods for lead optimization. Compared to traditional experimental high-throughput screening (HTS), virtual screening (VS) represents a more direct and rational approach to drug discovery, offering the advantages of low-cost and effective screening [7–9]. VS can be categorized into ligand-based and structure-based methods. Ligand-based methods, such as pharmacophore modeling and quantitative structure-activity relationship (QSAR) methods, are employed when a set of active ligand molecules is known, and little or no structural information is available for the targets [10]. On the other hand, structure-based methods rely on information derived from the 3D structure of a target of interest, allowing the ranking of molecule databases based on the structural and electronic complementarity of ligands to a given target [5].

In this context, molecular docking stands out as one of the most popular and successful structure-based in silico methods. It aids in predicting interactions between molecules and biological targets [5]. The process generally involves predicting the molecular orientation of a ligand within a receptor and then assessing their complementarity using a scoring function [5]. Since its initial emergence in the mid-1970s, docking has proven to be a crucial tool for comprehending how chemical compounds interact with their molecular targets, aiding in drug discovery and development. In fact, the number of studies reporting both the utilization of molecular docking to identify structural determinants essential for efficient ligand-receptor binding and the advancement of more precise docking methods has significantly increased since its inception [5, 10–24]. One of the earliest and most noteworthy studies on the application of docking in drug discovery and biology was conducted by Kuntz et al. in the early 1980s [10]. In this study, the authors presented a computational method that allowed the exploration of geometrically feasible ligand-receptor alignments for known heme-myoglobin/metmyoglobin and thyroxine/prealbumin structures [10]. While not the first study to employ docking for predicting potential conformations of molecular complexes [12], it marked the first use of a simplified function containing solely "hard sphere repulsions" and "hydrogen bonding" terms to describe protein-ligand interactions, which differed significantly from previous studies [12, 14, 15, 24]. Additionally, the authors were the first to consider the receptor as a solid rigid body with a binding site constituted by "pockets." Notably, the method in this study successfully predicted structures close to those of previously reported X-ray complexes and identified protein conformations suitable for energy refinement and potential ligand design [10].

However, the application of docking in drug design is constrained to biological targets with known crystal structures. Various approaches have been employed to overcome this limitation, such as constructing homology models derived from structural templates with highly homologous sequences when 3D structures are unavailable. Furthermore, these methods can be used in conjunction with molecular dynamics (MD) to validate and refine in silico modeled complexes [25–27]. Nevertheless, recent advancements in structural biology and crystal structure determination, progressively increasing access to experimentally derived ligand-target complexes [28–31], will undoubtedly alleviate this issue. In silico strategies, including molecular dynamics, have extensively explored the conformational space of investigated targets, ligands, and ligand-target complexes, providing a more comprehensive understanding of the dynamic behavior of ligand-target complexes and refining docking results [32, 33]. More rigorous virtual screening methodologies have also been developed to enhance docking-based predictions of ligand-target

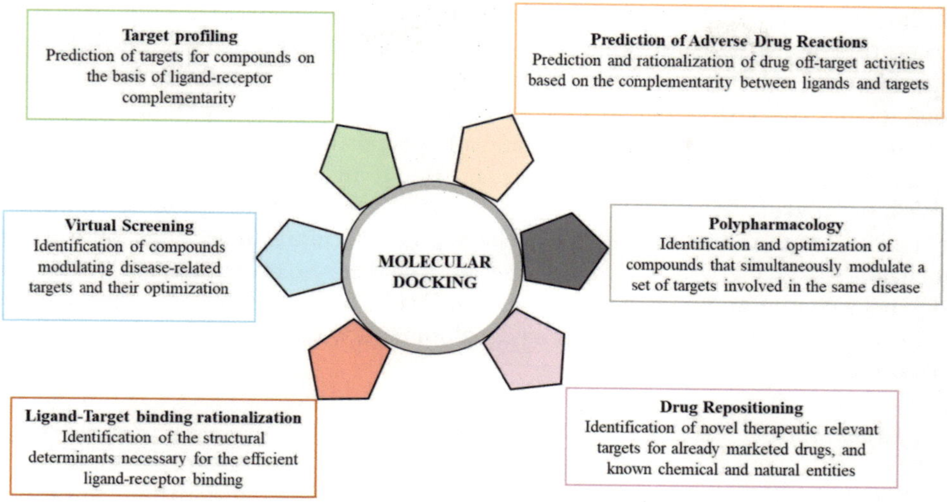

Figure 1.
Main applications of molecular docking in current drug discovery. Molecular docking is currently employed to help rationalize ligands activity toward a target of interest and to perform structure-based virtual screening campaigns, similarly to as when it was first developed. Besides these applications, it can also be used to identify series of targets for which the ligands present good complementarity (target fishing and profiling), some of them being potentially responsible for unexpected drug adverse reactions (off-targets prediction). Moreover, docking is also currently employed for the identification of ligands that simultaneously bind to a pool of selected targets of interest (polypharmacology) and for identifying novel uses for chemical compounds with already optimized safety profiles (drug repositioning).

complexes [34–38]. Indeed, these post-docking refinement and rescoring methods hold great significance in drug discovery, as they typically yield higher hit rates in virtual screening campaigns and demonstrate improved correlation with experimental data [36, 38]. Several reviews have addressed the role and applications of docking, exploring the potential it holds in drug design and development [39–42]. It is important to note that the uses and applications of docking have evolved since its inception. While initially designed to investigate molecular recognition between large and small molecules, docking is now extensively employed in various aspects of drug discovery programs. These include hit identification and optimization, drug repositioning, a posteriori target identification (reverse screening), multi-target ligand design, and repositioning (see **Figure 1**) [43–50].

Furthermore, docking facilitates the comprehension of relationships between different molecular targets associated with a specific disease, a crucial aspect for polypharmacology [51] and modern drug discovery in general.

In this chapter, we will discuss how various docking methods have been used to help assisting drug discovery tasks, giving particular emphasis on choosing the right software and docking protocols including challenges and limitations.

2. Fundamentals of molecular docking

2.1 Molecular docking defined

Molecular docking serves as a crucial tool in both structural molecular biology and computer-assisted drug design. Its primary objective is to forecast the primary binding mode(s) between a ligand and a protein with a known three-dimensional structure. The success of docking methods relies on their ability to effectively explore high-dimensional spaces and employ a scoring function that accurately prioritizes potential dockings. Utilizing docking enables virtual screening of extensive compound libraries, facilitating result ranking and proposing structural hypotheses regarding how ligands hinder the target—making it invaluable for lead optimization. It is essential to emphasize that the configuration of input structures for docking is as significant as the docking process itself [52]. Furthermore, analyzing results from stochastic search methods may sometimes pose challenges. In essence, molecular docking, a computational technique, predicts ligand-receptor binding affinity and has evolved into a formidable asset in drug development, with potential applications in nutraceutical research [53].

2.1.1 Explanation of molecular docking in computational biology

Completing the human genome project has uncovered numerous therapeutic targets for drug discovery. Advances in protein purification and structural analysis techniques, such as crystallography and nuclear magnetic resonance spectroscopy, have facilitated the integration of computational strategies in drug discovery. Virtual screening, a cost-effective and efficient approach, includes ligand-based and structure-based methods. Ligand-based methods, like pharmacophore modeling, are used when target structural information is limited [54] Molecular docking, a prominent method in structure-based drug design, models atomic-level interactions between small molecules and proteins, aiding in understanding biochemical processes. Efficient docking is achieved when the binding site is known through prior knowledge or identification using cavity detection programs. Early theories, like Fischer's lock-and-key and Koshland's induced fit, influenced molecular docking approaches [20]. Despite computer resource limitations, flexible ligand and rigid receptor docking remain popular, with ongoing efforts to address receptor flexibility challenges, including the proposed Local Move Monte Carlo (LMMC) approach [10].

2.1.2 Historical perspective

Molecular docking is a computer method utilized in drug development and structural biology. It includes predicting a molecule's preferred shape or orientation to form a stable complex bound to a target receptor. The historical perspective of this approach spans multiple decades:

1. *Emergence of Computational Chemistry (1960s–1970s)*: the 1960s and 1970s saw the emergence of computational chemistry. Molecular docking originated in the early years of computational chemistry when scientists started modeling and simulating molecular interactions using computers. The development of algorithms and the accessibility of computers provided the groundwork for the use of computational methods in the study of molecular interactions and structures [55]

2. *Invention of Molecular Mechanics (1970s–1980s)*: more accurate simulations of molecular interactions were made possible by the invention of molecular mechanics force fields, which characterize the energy and forces in a molecular system. These force fields were first used by scientists to simulate how molecules would behave in three dimensions [56]

3. *The Advent of Molecular Dynamics (1970s–1980s)*: this dynamic perspective on molecular interactions was made possible by molecular dynamics simulations, which imitate the real-time movements of atoms and molecules. Researchers were able to watch how molecules behaved and bound to receptors over time thanks to this technology [55, 56]

4. *Pioneering Work (1980s–1990s) on Docking Algorithms*: in the 1980s and 1990s, few researchers contributed to the creation of the first molecular docking algorithms. The initial docking algorithms relied on basic geometric and empirical scoring functions to anticipate the binding affinity and alignment of ligands within the target protein's binding region [57].

5. *Scoring Function Advancements (1990s–2000s)*: with the advent of increasingly complex scoring functions, docking algorithms saw notable advancements in the 1990s and 2000s. The purpose of these scoring functions was to differentiate between genuine binding poses and false positives, as well as to better anticipate the binding free energy [55].

6. *Use in Drug Discovery (2000s–Present)*: in the pharmaceutical sector, molecular docking is now a crucial component of drug discovery workflows. Large compound libraries are screened using it to find possible therapeutic candidates, forecast binding affinities, and enhance lead compounds.

7. *Machine Learning Advancements (2010s-Present)*: the accuracy of molecular docking predictions has been significantly improved in recent years by including machine learning techniques, such as deep learning. These methods enhance our comprehension of intricate molecular interactions by utilizing extensive datasets and advanced algorithms [58].

The advancement of computational methods and the incorporation of experimental data to enhance comprehension and forecast molecular interactions for diverse uses, most notably drug discovery, are reflected in the historical development of molecular docking. Artificial intelligence (AI) has been applied to various docking systems in different domains. In the field of cross-docking systems, AI-based techniques have been explored to solve cross-docking problems, and new potential uses of AI-based techniques have been identified [59]. In the context of subsea homing and docking systems for autonomous underwater vehicles (AUVs), an AI-enabled short-range AUV electromagnetic homing guidance system (EMHGS) has been developed using supervised and unsupervised machine learning algorithms [60]. In the domain of protein-ligand docking, high-speed ML models have been used to accelerate the scoring of compounds, and AI-based models have been utilized as a pre-filter to improve the speed and accuracy of virtual screening [61]. In the field of on-orbit servicing (OOS), an AI-powered navigation algorithm has been developed for autonomous docking and refueling of spacecraft [62]. In the domain of underwater docking, deep learning techniques, including knowledge distillation and synthetic data generation, have been used to efficiently train convolutional neural networks (CNNs) for docking station detection [63].

2.2 Biological relevance

Molecular docking is a computational method employed to identify the correct binding pose of a protein-ligand complex and assess its strength through various scoring functions, ranking the best pose for each molecule [10]. The technique aims to fit a ligand into the target protein's binding site by optimizing variables like hydrophobic, steric, and electrostatic complementarity, estimating their binding free energy. Successful docking relies on high-resolution structures, obtained through X-ray crystallography, NMR, or homology modeling, with a known binding site.

The Protein Data Bank (PDB) contains thousands of experimentally determined 3D structures of proteins. X-ray and NMR machines enable the detailed resolution of molecular interactions, crucial for recognizing key residues, analyzing interaction forces, and energetics, and understanding structural fits. Some protein structures are challenging to determine experimentally, leading to their absence in the PDB. Yet, advancements in bioinformatics, like homology modeling, enable the prediction of unavailable protein structures using templates with similar residues. Servers include Swiss Model, Phyre [58] i-TASSER, HH Pred, PSI-Pred, Robetta, Raptor, and the groundbreaking AlphaFold by DeepMind [58]. It has revolutionized computational structural biology in 2021, achieving over 92% accuracy in 3D structural prediction [64]. AlphaFold and RoseTTAFold, another accurate model, hold promise for transforming in silico drug discovery in the twenty-first century, with RoseTTAFold excelling in predicting complex biological assemblies like protein-protein interactions [64].

2.3 Types of molecular docking

Molecular docking is a valuable tool in drug discovery research because it makes target selection, validation, virtual lead identification screening, and lead optimization easier. Drugs are primarily organic tiny molecules; however, targets might be either protein or DNA depending on the type of sickness of concern. To accomplish these objectives, many molecular docking approaches are used, such as protein-protein, protein-DNA, protein-small molecule, and DNA-small molecule. Majorly docking can be classified into two categories:

- *Flexible docking*: addressing flexibility in molecular docking poses a significant challenge in cell biology research. The flexible docking process comprises four main stages. Firstly, in the preprocessing stage, proteins undergo analysis to define their conformational space, allowing the generation of discrete conformations. This ensemble is used in cross-docking, simulating the conformational selection model. Hinge locations may also be identified, leading to the separation of rigid and flexible parts for individual docking [8]. The second stage involves rigid docking, aiming to produce solution candidates with at least one near-native structure, permitting some steric clashes. The subsequent refinement stage models induced fit, optimizing each candidate through small backbone and side-chain movements and rigid-body adjustments. This optimization is performed in successive steps for side-chain conformations, backbone structure, and rigid-body orientation. The refined structures exhibit improved binding energy and minimal steric clashes. The final scoring stage assesses and ranks candidate solutions based on parameters like binding energy, agreement with known binding sites, deformation energy of flexible proteins, and the presence of energy funnels [65].

- *Rigid docking*: rigid docking is a stage in molecular docking where the docking procedure is conducted without allowing significant flexibility or movement in the structures of the molecules being docked. In other words, the interacting molecules, typically a ligand and a receptor (e.g., protein), are treated as rigid bodies during the docking simulation. This approach assumes that the overall structure of the molecules remains fixed, and only translational and rotational degrees of freedom are considered in the docking process [10]. During rigid docking, the goal is to generate a set of potential docking poses for the ligand within the binding site of the receptor. This stage aims to identify configurations where the ligand and receptor interact favorably, forming a stable complex. While the rigid docking stage allows for some steric clashes, it does not consider significant conformational changes or flexibility in the molecular structures [66]. Rigid docking is often followed by a refinement stage, where more detailed adjustments, including flexible movements of side chains or larger conformational changes, may be considered to improve the accuracy of the predicted binding poses. Employing rigid or flexible docking depends on the specific goals of the docking study and the nature of the interactions being investigated [67].

2.4 Other variations and hybrid methods

In molecular docking, hybrid methods pertain to the amalgamation of many computational methodologies or approaches to enhance the precision and dependability of forecasts. These hybrid approaches make up for the shortcomings of individual techniques by combining the advantages of several methodologies. The following are a few typical hybrid molecular docking technique types:

2.4.1 Hybrid techniques based on structure and ligands

Pharmacophore-Based Docking: combines structure-based docking and ligand-based pharmacophore modeling. New compounds are docked using pharmacophore properties that come from well-known ligands [5].

Shape-Based Docking: integrates structure-based docking with knowledge of the shape of the ligand. The shape similarity between ligands and the target binding site is considered during the docking process [6].

2.4.2 Hybrid docking methods and molecular dynamics (MD)

The docking procedure is guided by data obtained from molecular dynamics simulations. The accuracy of docking poses can be improved, and conformational changes observed during MD simulations can be used as guidance using molecular dynamics simulations on anticipated docking poses, MD-refined docking involves fine-tuning and optimizing the complicated structures of ligands and receptors [6].

2.4.3 Docking and quantum mechanics/molecular mechanics (QM/MM)

QM/MM-docking combines molecular mechanics for the remainder of the system with quantum mechanics for precisely handling sections (such as the active site). Studying processes catalyzed by enzymes is an excellent use for this hybrid approach.

2.4.4 Docking and machine learning: ML-based scoring functions

Combines machine learning models to enhance the scoring functions employed in docking investigations. With the help of experimental data, machine learning algorithms can more precisely estimate binding affinities. ML-enhanced pose prediction: combines machine learning techniques with docking algorithms to improve pose prediction accuracy, especially for flexible ligands and dynamic binding sites.

2.4.5 Ensemble docking

Multiple receptor conformations: involves using an ensemble of receptor structures (e.g., from different conformations or experimental structures) during the docking process. This approach considers the inherent flexibility of the receptor.

2.4.6 Hybrid ligand screening

It improves the discovery of putative ligands from extensive compound libraries by fusing structure-based docking with ligand-based virtual screening.

2.4.7 Docking and free energy calculations

To estimate the binding free energy of ligands more precisely, free energy perturbation (FEP) docking integrates docking with free energy calculations.

These hybrid approaches seek to overcome issues with molecular docking scoring accuracy, ligand flexibility, and receptor flexibility. Enhancing the efficacy of molecular docking in drug discovery and structural biology, researchers can attain more dependable predictions of ligand binding mechanisms and affinities by merging complementing techniques.

3. Computational tools and software

A vital tool in the drug-discovery process, docking software helps scientists understand how small compounds interact with proteins of interest, leading to the creation of novel therapeutic medicines. To predict the ideal binding configuration. Three areas use computational tools: virtual screening to find compounds with high binding potential through large databases, lead optimization to improve lead compound properties, and drug discovery to find potential drug candidates by predicting their binding affinity to target proteins [68, 69].

3.1 Overview of docking software

Docking software predicts the binding mechanism and affinity of small compounds with a target biomolecule, usually a protein, and is therefore essential to computational chemistry and drug discovery [70].

3.2 AutoDock Vina

AutoDock Vina is a molecular docking software used in computational chemistry to predict the binding modes of small molecules or ligands with a target protein. It is an improved version of the original AutoDock software and is widely used for virtual screening and drug discovery studies [70].

Key features of AutoDock Vina include:

Scoring function: AutoDock Vina uses an advanced scoring function to evaluate the binding affinity between a ligand and a target protein. The scoring function takes into account various energy terms including steric clashes, hydrogen bonding, and electrostatic interactions.

Search algorithm: it employs a global optimization algorithm called iterated local search (ILS) to explore the conformational space efficiency and find energetically favorable binding modes.

Flexible ligand and receptor handling: AutoDock Vina allows for the flexibility of both ligands and receptors. This is important as proteins and ligands can undergo conformational changes upon binding.

User-friendly interface: while AutoDock Vina itself is a command line tool, there are graphical user interfaces (GUIs) available that provide a more user-friendly environment for setting up and running docking simulations.

Open source: AutoDock Vina is open-source software, making it freely available for academic and research purposes. Users can access the source code, modify it, and contribute to its development.

Integration with visualization tools: results from AutoDock Vina can be visualized using molecular visualization tools, such as PyMOL or visual molecular dynamics (VMD), to analyze the binding modes and interactions between ligands and proteins. It is important to note that AutoDock Vina is one of many molecular docking tools available, and the choice of which tool to use depends on the specific requirements of the research or virtual screening project. Users are encouraged to refer to the official documentation and publications associated with AutoDock Vina for detailed information on its usage and capabilities [71].

3.2.1 Glide

A Schrodinger company known for its suite of computational chemistry and molecular modeling software does offer a docking program called Glide [19]. Glide is a molecular docking program used for predicting the binding modes and affinity of small molecules to target proteins. It is widely used in drug discovery and structure-based drug design. Glide is the primary industrial solution for reliable ligand-receptor docking. It improves and speeds up structure-based drug design for a variety of uses, such as interactive 3D molecular design, virtual screening, and binding mode prediction. It is easy to understand that a guided graphical user interface makes it simple to develop and test docking models and also achieve high enrichment for small compounds, peptides, and macrocycles among a wide variety of receptor types. Additionally, it offers simple, constraint-free docking computations that satisfy the criteria and desired chemical space as found in experiments. Schrödinger programs can be executed via a command line or an interface with graphics. The working directory is the location to which the software writes input and output files. The working directory for the task is the location from which you launch the application when using the command line.

The software package developed by Schrodinger utilizes several methods and methodologies to facilitate activities associated with drug development, molecular modeling, and computational chemistry. The Glide algorithm, which is utilized for molecular docking, is one of the well-known algorithms in Schrodinger's program. The first stage of Glide, High Throughput Virtual Screening (HTVS), is intended for the quick screening of many ligands.

Standard Precision (SP): using a more precise scoring function and considering a smaller subset of ligands, this phase improves upon the findings from HTVS.

Extra Precision (XP): this Glide phase is the most precise and computationally intensive. Regarding ligand-receptor interaction prediction, it offers the highest degree of accuracy.

Schrodinger's software suite includes Glide and several other algorithms and tools for various applications, including pharmacophore modeling, quantum mechanics computations, and molecular dynamics simulations [72].

3.2.2 Genetic optimization for ligand docking (GOLD)

A genetic technique called Genetic Optimization for Ligand Docking (GOLD) is used to dock flexible ligands into protein binding sites [19, 20]. GOLD has undergone comprehensive testing and demonstrated outstanding performance in virtual screening and strong outcomes in pose prediction. GOLD optimizes ligand placement by exploring the conformational space of the ligands within the binding site using a genetic algorithm. The scoring function in GOLD is critical for measuring the fitness of different ligand conformations created during the docking process. Typical terms and components relevant to GOLD scoring functions in molecular docking software are as follows:

Vander Waals interactions explain the forces that attract and repel atoms. Electrostatic interactions consider the interactions between charged particles, such as those between positively and negatively charged atoms.

Hydrogen bonding: the creation of hydrogen bonds between the ligand and the protein may be considered a scoring function.

Solvation energy refers to the energy needed to dissolve the binding site and the ligand.

Entropic contributions: after binding, entropy changes may also be considered by scoring functions.

Flexibility: by permitting ligand flexibility or considering protein-induced-fit effects, several scoring methods consider both the ligand and the protein's flexibility [73].

3.2.3 Moledock/Molegro

Molegro Virtual Docker is an integrated platform for predicting protein-ligand interactions, from preparing the compounds to identifying possible target protein binding sites and forecasting ligand binding modalities, Molegro Virtual Docker manages every step of the docking process [74].

Molegro Virtual Docker is a user interface that prioritizes productivity and usability while providing high-quality docking based on a unique optimization technique.

High docking accuracy: it has been demonstrated that the docking engine can accurately recognize binding modes. It has been demonstrated that Molegro Virtual Docker performs better than other docking programs when it comes to identifying the proper binding modes.

Easy-to-use interface: the integrated wizards simplify the process of setting up and executing docking runs for the user. Sophisticated visualization and analysis tools are provided to study ligand-receptor interactions and optimize found docking solutions [75].

3.2.4 FlexX

One important tool for assessing and ranking the possible binding positions of ligands within a target binding site is the scoring function included in molecular docking software, such as FlexX [76, 77]. To assist researchers in determining the most likely and biologically significant binding modes, the scoring function attempts to estimate the binding affinity between the ligand and the receptor. One important tool for assessing and ranking the possible binding positions of ligands within a target binding site is the scoring function included in FlexX. To assist researchers in determining the most likely and biologically significant binding modes, the scoring function attempts to estimate the binding affinity between the ligand and the receptor. FlexX evaluates a docking pose's quality using a scoring algorithm that considers several variables [78].

Ligand-receptor interaction energy: this metric assesses how strongly a ligand and receptor interact. Van der Waals forces, hydrogen bonds, and electrostatic interactions are examples of this.

Desolvation energy: the cost of desolvating the ligand and receptor as they assemble in the binding site is expressed in terms of desolvation energy.

Entropy-related terms: certain scoring functions take conformational entropy or loss of freedom into account when calculating scores.

Hydrophobic interactions: the scoring function may additionally take hydrophobic contact contributions into account.

Additional structural and energetic features: additional terminology about receptor flexibility, ligand flexibility, and other structural or energetic qualities may be provided, depending on the particular program.

3.2.5 Surflex-Dock

A molecular docking tool called Surflex-Dock is used to forecast the affinities and binding patterns of ligands to protein targets [78]. The Hammerhead scoring function, which considers both form complementarity and electrostatic interactions between ligands and proteins, serves as the foundation for the scoring system in Surflex-Dock. The fitness of a ligand posture inside a protein binding site is assessed using the scoring function [79].

Shape complementarity (Sshape): measure how well the ligand fits into the protein's binding site according to shape.

Chemical score (Schem): analyze the chemical complementarity of the ligand and protein using the chemical score. Ligands with advantageous chemical interactions obtain better scores.

Solvation (Ssolve): explains the change in salvation energy upon ligand binding and encourages ligands that displace water molecules from the binding site.

Torsional Strain (Stors): penalizes strained torsional angles in the ligand. Lower scores are assigned to ligand poses with unfavorable torsional geometry.

3.2.6 ICM-pro

ICM-Pro is a molecular docking tool, and user guides and documentation are available from the program's creators or the official Web site. With direct access to sequence and structural databases, the docking software ICM-Pro expands the capabilities available to biologists and chemists. User can perform the following tasks: generate surfaces, compute electrostatics, introduce mutations, analyze sequences and alignments, look

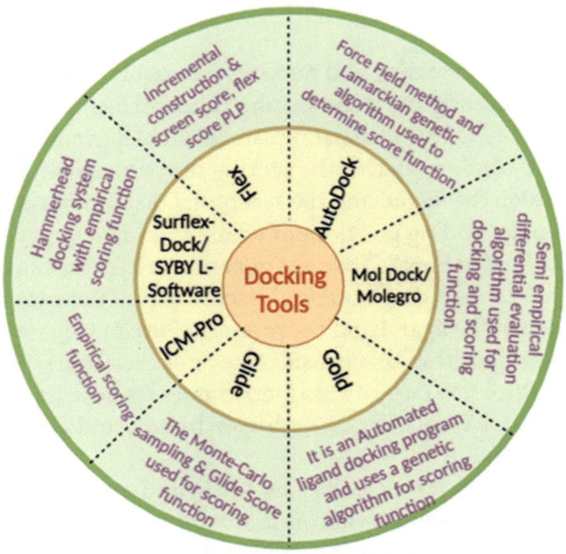

Figure 2.
Different docking tools available.

at protein structure, look at pockets and bound ligands and medications, and dock tiny molecules and proteins with one another. Determine the locations of ligand binding sites and create ligands with the Interactive Ligand Editor (**Figure 2**) [71, 75].

3.3 Choosing the right software

Choosing the molecular docking software depends upon the type of work or research you are doing, your budget, your level of experience, and your unique needs will all play a role in selecting the best molecular software. When choosing molecular software, keep the following important factors in mind.

3.3.1 Considerations for selecting docking software

- User interface (UI) and user-friendliness
- Integration and compatibility
- Compliance and data security
- Trial period
- Cost and licensing
- Compliance and data security
- Features and functionality.

A user interface that is easy to use is essential, especially if your team is diverse and comprises members with varying skill sets. Look for software that has a simple user interface, clear navigation, and comprehensive help. Check to see if the software works with the operating system you are using Windows, macOS, or Linux. Consider the program's compatibility with databases and other applications that you may find yourself using. If you work with sensitive data or are required to follow particular instructions, make sure the program conforms with security standards and legislation. Use trial copies or demos of the program to determine how well it fits into your particular workflow before making a final choice. Regardless matter whether the product is open source, commercial, or freeware, find out the license structure. Keeping in mind your budgetary limitations, select a program that fits within what you can afford. If the application must follow certain criteria or you work with important information, make sure it conforms with security norms and regulations. Verify that the program includes the exact features and capabilities you need for your research or task. This could involve docking investigations, dynamics simulations, simulation, visualization, and molecular modeling, among other things [80, 81].

3.3.2 Preparation of ligands and receptors

This section will guide you through the essential steps involved in preparing receptors and ligands for molecular docking experiments, ensuring accurate and reliable results in computational studies.

3.3.3 Preparation of ligand

A key initial step in gaining significant and physiologically relevant information from molecular docking studies is to ensure the structural quality of ligands. To improve the accuracy of the docking predictions, researchers should carefully validate and optimize ligand structures.

3.3.3.1 Retrieval of structure

The three-dimensional structures of ligand obtained from the software like ChemDraw or Marvin Sketch, or you may utilize chemical databases like PubChem and ChEMBL.

3.3.3.2 Conversion into desired file format

Convert ligand structures into a format that docking software recognizes. Protein Data Bank (PDB) and molecular 2 (mol2) formats are common formats.

3.3.3.3 Minimizing energy

Minimize the ligands' energy to get them into stable, low-energy conformations. This stage contributes to increasing the correctness of the docking findings. One can utilize programs such as PyMOL, Open Babel, or Avogadro.

3.3.3.4 Protonation and ionization

Reduce the ligands' energy and transform them into stable, low-energy conformations. This step helps to improve the accuracy of the docking findings. One can use applications like Avogadro, PyMOL, and Open Babel.

3.3.4 Preparation of receptors

There are multiple phases involved in the production of receptors in the context of molecular docking to make sure that the protein structure is appropriate for docking investigations. A computer forecasts the manner and affinity of a ligand's (small molecule's) binding to a target protein. The following are general instructions for setting up receptors in docking software, along with links to other resources:

3.3.4.1 Protein structure retrieval

Acquire the receptor protein's three-dimensional structure. It is possible to employ homology modeling, NMR spectroscopy, or X-ray crystallography.

3.3.4.2 File format conversion

The second step is to convert the protein structure into a file format that docking software can read. PDB format is widely utilized.

3.3.4.3 Removal of water and non-biological molecules

Remove the protein structure of any water molecules and non-biological molecules (such as ions or co-crystallized ligands). Chimera and PyMOL are two tools that can help with this process.

3.3.4.4 Add hydrogen

Verify that the hydrogen atoms are positioned correctly in the protein structure. Accurate docking predictions depend on this. To add hydrogen atoms, using programs such as PDB2PQR or Reduce.

3.3.4.5 Energy minimization

To reduce any steric conflicts and improve the geometry, minimize the energy in the protein structure. Software for molecular dynamics or programs like CHARMM or GROMACS can be utilized.

3.3.4.6 Active site positioning

Determine and describe the protein's binding pocket, also known as the active site, where the ligand is predicted to bind. This may be supported by computational predictions, literature, or experimental results.

3.3.5 Importance of structural quality

In molecular docking investigations, the structural quality of ligands is important since it can greatly affect the precision and dependability of the docking data. The following, along with relevant references, are some salient points highlighting the significance of ligand structural quality in molecular docking:

3.3.5.1 The accuracy of binding prediction

The importance of precisely predicting binding conformations and energies is accurate ligand structures are required.

3.3.5.2 Energy minimization and conformational sampling

Improved ligand architectures enhance reliability in energy minimization and improved conformational sequencing during the docking procedure.

3.3.5.3 Handling tautomers and ionization states

Ligand structures must accurately depict the relevant ionization and tautomeric states in the context of the experiment.

3.3.5.4 Handling ionization states and tautomers

Ligand structures ought to faithfully depict the pertinent ionization and tautomeric states in the context of the experiment.

3.3.5.5 The ionization states and pH sensitivity

Based on the experiment's results, pH should be considered, as ligand protonation states impact binding affinity.

3.3.5.6 Handling ligand flexibility

Accurate ligand structures are essential for managing ligand flexibility, an important factor in several drug-receptor interactions.

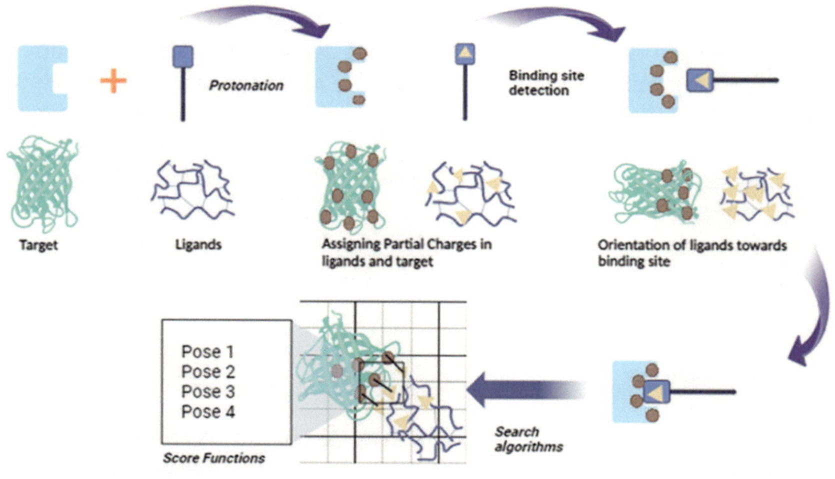

Figure 3.
Process of molecular docking.

3.3.5.7 Docking software considerations

Different ligand structures may be suggested or required by different docking software.

3.3.5.8 Validation and benchmarking

Assessing experimental findings against ligand structures validates them and improves the validity of docking investigations (**Figure 3**).

4. Docking protocols

4.1 Pre-docking steps

It is essential to synthesize and purify ligand and receptor molecules before using them in different biochemical and biophysical research projects. Small chemical molecules called ligands attach to particular receptors with preference. Many biological functions, such as signal transduction, cellular communication, and therapeutic efficacy, depend on this interaction between ligands and receptors. The ligand structure can be obtained from databases like ZINC [77] and Pub Chem (https://pubchem.ncbi.nlm.nih.gov/) or sketched using the Chem sketch tool. The ligand can be screened using the Lipinsky Rule of Five as a pre-docking phase for ligand preparation [82].

The Protein Data Bank (PDB) has the protein's three-dimensional structure available for download. This recovered structure needs to go through pre-processing after that. According to the given parameters, this pre-processing entails, among other things, eliminating water molecules from the protein cavity, stabilizing charges, adding missing residues, and producing side chains [83].

4.1.1 Grid generation

The area surrounding a protein molecule's binding site in three dimensions is commonly referred to as a grid within molecular docking processes. Various tools, such as CASTp (http://sts.bioengr.uic.edu/castp/) and LigASite A (http://www.bigre.ulb.ac.be/Users/benoit/LigASite/index.php?home), are available for predicting an active site. This grid, divided into evenly spaced points around the active site, essentially represents a discretized version of the binding site. Each point on the grid represents a potential location for a ligand molecule to bind to the protein. Assessing the protein's active site is essential. While the receptor may possess multiple active sites, only the one presenting the highest risk must be prioritized for selection.

4.2 Docking algorithms

Docking algorithms play a crucial role in drug discovery and molecular design by predicting how small molecules (ligands) bind to larger molecules (proteins or receptors). The docking algorithms aim to determine which conformation is the most stable and energetically favorable and what forms the ligand attached to the protein could take.

4.2.1 Different docking approach

- *Shape-based docking:* focuses on the binding site's and ligand's geometric compatibility. Examples of software falling in this category include DOCK, which employs geometric matching algorithms to identify poses with optimal shape complementarity between ligand and receptor. FTDOCK utilizes the fast Fourier transform to compare the shapes of ligands and receptors efficiently. ZDOCK leverages molecular surface patches to rapidly assess geometric compatibility.

- *Docking with energy grids:* potential energy at each location in a discretized binding site grid is calculated throughout the docking process using energy grids. Examples of software falling in this category include AutoDock, which calculates intermolecular interaction energies (e.g., van der Waals, electrostatic) on a pre-calculated grid around the binding site. Glide [67] employs a grid-based scoring function combined with ligand conformational sampling to identify energetically favorable poses. GOLD [10] utilizes a pre-calculated potential field grid to estimate interaction energies and guide ligand placement.

- *Knowledge-based docking:* makes more precise predictions using data from known protein-ligand complexes. Examples of software in this category include DFIRE, scores docked poses based on their similarity to known protein-ligand complexes extracted from structural databases. ICM-Pro leverages knowledge-based potentials derived from protein-ligand interfaces to rank potential binding modes. RosettaDock [84] integrates structural information from protein-protein interfaces to evaluate docked poses (**Table 1**).

S. no	Algorithm	Software	Principle	Scoring function	References
1.	Lamarckian Genetic Algorithm (LGA)	AutoDock and AutoDock Vina	LGA combines aspects of genetic algorithms with local search methods. It evolves a population of candidate solutions and applies local search operators to improve solution quality.	Empirical scoring functions based on intermolecular interactions such as van der Waals forces, electrostatic interactions, and hydrogen bonding.	[56]
2.	Genetic Algorithm (GA)	GOLD and AutoDock	GA is a search heuristic inspired by natural selection and genetics principles. It maintains a population of candidate solutions and applies genetic operators such as selection, crossover, and mutation to evolve toward optimal solutions.	GA utilizes empirical scoring functions to evaluate the fitness of candidate solutions based on intermolecular interactions.	[10, 56]
3.	Differential Evolution Algorithm (DE)	AutoDock and AutoDock Vina	DE is a population-based optimization algorithm that iteratively improves candidate solutions by combining differences between randomly selected solutions.	Employs scoring functions that evaluate the energy of the docking complex, considering factors like steric clashes, hydrogen bonding, and electrostatic interactions.	[56]

S. no	Algorithm	Software	Principle	Scoring function	References
4	Monte Carlo (MC) methods	Rosetta, FlexX	MC methods use random sampling techniques to estimate properties of a system by generating a large number of random configurations and averaging their contributions.	MC methods utilize scoring functions that assess the energy of the system based on molecular mechanics force fields or empirical potentials.	[84]
5. 5	Ant Colony Optimization (ACO)	SwarmDock	The foraging behavior of ants inspires ACO. It utilizes pheromone trails left by ants to guide the search for optimal solutions in a graph-based problem space.	ACO relies on scoring functions that evaluate the fitness of candidate solutions by considering the complementarity of molecular surfaces and the strength of intermolecular interactions.	[85]
6.	Particle Swarm Optimization (PSO)	eHiTS and SwarmDock	PSO is inspired by the social behavior of bird flocking and fish schooling. It maintains a population of candidate solutions (particles) that move through the search space based on their best position and the global best position the swarm finds.	PSO utilizes scoring functions similar to those used in other optimization algorithms, assessing the fitness of candidate solutions based on the energy of the docking complex and the quality of molecular interactions.	[85]
7.	Tabu Search (TS)	FlexX	TS is a metaheuristic search method that iteratively explores the solution space by maintaining a tabu list of recently visited solutions and applying neighborhood search techniques to escape local optima.	TS employs scoring functions that evaluate the energy and stability of candidate docking configurations, often based on empirical potentials or molecular mechanics force fields.	[86]
8.	Simulated Annealing (SA)	Rosetta and FlexX	The annealing process inspires SA in metallurgy. It probabilistically accepts worse solutions in the early stages of optimization and gradually decreases the probability as the algorithm progresses.	SA utilizes scoring functions that evaluate the energy of the system and guide the search for low-energy conformations based on molecular mechanics force fields or empirical potentials.	[84, 85]
9.	Iterated Local Search (ILS)	PatchDock, SwarmDock	ILS iteratively explores the solution space by perturbing current solutions and applying local search techniques to improve solution quality.	ILS employs scoring functions that assess the fitness of candidate solutions based on the energy and stability of the docking complex, considering factors like steric clashes and intermolecular interactions.	[86, 87]

S. no	Algorithm	Software	Principle	Scoring function	References
10.	Evolutionary Strategies (ES)	Rosetta and AutoDock	ES is a family of optimization algorithms inspired by biological evolution. It utilizes mutation, recombination, and selection to evolve a population of candidate solutions toward optimal solutions iteratively.	ES typically utilizes scoring functions similar to those used in other optimization algorithms, evaluating the fitness of candidate solutions based on the energy and stability of the docking complex.	[56, 84]

Table 1.
Outline of specific software applications in docking and their primary principles.

4.3 Post-docking analysis

4.3.1 Scoring functions

Scoring functions are crucial in molecular docking studies to evaluate the quality and feasibility of anticipated ligand-receptor interactions. These functions aim to characterize the binding affinity between a ligand and a receptor by evaluating various energy contributions [83]. A scoring function consists of terms related to van der Waals forces, electrostatic interactions, hydrogen bonding, dissolution effects, and sometimes additional terms for specific interactions. The docking algorithm's various ligand binding postures are ranked and prioritized using the scoring function during post-docking analysis. Finding the most energetically advantageous binding modes is the aim. Scoring functions are estimates of the actual binding energy and might not adequately represent all facets of molecular interactions, and thus, it is essential to understand their limits. In order to direct the optimization process and find energetically advantageous docking configurations, scoring functions are essential. A brief overview of the scoring function employed by various docking techniques is provided in **Table 2**.

4.3.2 Visualization tools

In molecular docking studies, visualization methods are essential for analyzing and understanding the complicated three-dimensional structures of ligand-receptor complexes. These tools allow scientists to gain additional insight into molecular interactions' general geometry, interactions, and binding mechanisms. Examples of frequently used visualization tools are Pymol, UCSF Chimera, JMOL, and RASMOL. While it is outside the scope of this chapter to cover all the tools, several excellent references provide information on visualization software.

5. Challenges and limitations

Common challenges and pitfalls in molecular docking include:

Protein flexibility: ignoring protein flexibility can lead to inaccurate predictions as proteins can undergo conformational changes upon ligand binding. Incorporating protein flexibility in docking algorithms is crucial to account for induced fit effects.

Software's	Designer company and year of introduction	Docking approach/ scoring function	Relevancy and novelity of softwares	References
AutoDock	D. S. Goodsell and A. J. Olson The Scripps Research Institute 2009	It is based on force field methods and follows the Lamarckian genetic algorithm used for scoring function.	Open-source software, improved accuracy and speed, and user-friendly interface	[70]
Glide	Schrodinger Inc. 2022	It is based on Monte Carlo sampling & GlideScore used for scoring function	Ligand-based design quantum mechanics molecular dynamics simulations	[72]
Gold	Cambridge Crystallographic Data Centre 1997	It is based on automated ligand docking program that uses a genetic algorithm for scoring function	Based on genetic algorithm & six different scoring functions	[88]
Mol Dock/ Molegro	Molexus a Danish company 2005–2008	It is based on semi-empirical differential evaluation algorithm	High docking accuracy, easy-to-use interface, and proprietary scripting languages	[73]
Flex X	T. Lengaurer and M. Rarely BiosolvetT 2023	It is based on incremental construction & screen score, flex score PLP	Based on Monte Carlo algorithm & performed templet docking covalent docking (structure-based docking)	[89]
Surflex-Dock/SYBYL software 1990	BioPharmic	Hammerhead docking system having empirical scoring function	Show different potential binding modes, advanced feature	[90]
ICM-Pro	Molsoft LLCI 1985	Emperical	Monte Carlo algorithm, with crystallographic analytical tools	[71, 75]

Table 2.
Summarizes some commonly used protein-ligand docking tools.

Active site water molecules: handling active site water molecules can be challenging. Incorrect treatment of water molecules may introduce false energetics and affect the accuracy of docking results.

Scoring function selection: choosing an inappropriate scoring function can impact ligand ranking and pose selection. Using scoring functions suitable for the specific docking task and target is essential.

Ligand conformational sampling: inadequate sampling of ligand conformations can result in missing potential binding modes. Proper ligand conformational sampling techniques are necessary to explore the full conformational space of ligands.

Structural data quality: reliance on low-quality or incomplete structural data for proteins and ligands can lead to unreliable docking results. Using high-quality structural data from reputable sources like the PDB is crucial for accurate predictions.

DOI: http://dx.doi.org/10.5772/intechopen.1005527

Algorithm selection: choosing the wrong docking algorithm for a specific target or ligand can result in suboptimal docking performance. It is important to consider the characteristics of the target and ligands when selecting a docking method.

Overlooking protein-ligand interactions: failing to consider all relevant protein-ligand interactions, such as hydrogen bonding, hydrophobic interactions, and electrostatic interactions, can lead to inaccurate binding predictions.

Scoring function limitations: scoring functions may have limitations in accurately predicting binding affinities or distinguishing between active and inactive ligands. Understanding the strengths and weaknesses of different scoring functions is important for interpreting docking results.

6. Case studies

As invaluable insights into using computer methods for comprehending protein-ligand interactions and drug discovery in molecular docking. A handful of the research that is presented below demonstrates the molecular docking and virtual screening procedures that are used to look into viral proteins, create new vaccines, and assess possible treatments. These techniques provide crucial possibilities for addressing major global health issues.

6.1 Case study 1: chikungunya virus E glycoprotein

Chikungunya virus (CHIKV) outbreaks have been widespread across Asia, Africa, and Central and South America, causing significant public health concerns. Due to its positive sense RNA nature, CHIKV's replication within the host has led to mutations in its genome, posing challenges in vaccine development and treatment strategies. While vector control remains effective in curbing CHIKV transmission, eradicating the vector species presents formidable obstacles [91]. Numerous conserved epitopes across different CHIKV strains have been identified through extensive investigation. Leveraging these putative antigenic epitopes holds promise for developing innovative vaccines to combat and eradicate CHIKV infections.

6.2 Case study 2: dissection of capsid protein HPV

Human papillomavirus type 52 (HPV 52) stands as one of the most concerning strains contributing to cervical cancer globally. This study explored potential vaccination approaches against cervical cancer utilizing computational techniques such as immuno-informatics and molecular docking [92]. The research presents a robust analysis and practical methodology for efficiently generating vaccines.

6.3 Case study 3: in silico rational design of a novel tetra-epitope tetanus vaccine

Exposure to various epitopes within the tetanus toxoid vaccine can produce undesirable antibodies by lymphocyte clones. Ideally, tetanus vaccination should cater to individuals with specific human leukocyte antigen alleles. *Clostridium tetani* bacteria, responsible for tetanus, release toxins causing severe muscular contractions, often termed as "lockjaw." This study explores the rational design of a tetra-epitope tetanus vaccine, aiming to provide comprehensive protection against tetanus while considering genetic diversity [93].

6.4 Case study 4: evaluation of potential anti-hepatitis A virus

Hepatitis A virus (HAV) remains a common cause of acute hepatitis globally, occasionally leading to acute liver failure. Despite the availability of the HAV vaccine, its prevalence in certain regions persists. This research assesses a potent HAV 3C protease inhibitor effective against human hepatoma cells infected with HAV genotype IIIA HA11-1299. Highlighting the significance of hydroxyl groups in inhibiting protease activity, the study suggests small molecule inhibitors as a viable strategy for developing novel anti-HAV therapeutics [94].

6.5 Case study 5: monkeypox virus

Monkeypox, caused by a double-stranded DNA virus belonging to the Poxviridae family, shares structural similarities with the smallpox virus. Recent outbreaks of monkeypox in African nations pose a significant public health risk. Molecular docking studies revealed hydrophobic interactions between FDA-approved medications and the proteinase amino acid residues of the monkeypox virus. This research sheds light on potential therapeutic targets for managing monkeypox infections in public health emergencies [95].

7. Conclusions

In conclusion, the chapter encapsulates the foundational aspects and significant applications of molecular docking in contemporary research. Beginning with elucidating the fundamental principles underlying protein binding and activation, the chapter navigates through the protein-protein interactions and ligand binding mechanisms essential for drug development. By exploring the diverse array of software tools available for molecular docking, the chapter sheds light on the underlying algorithms that power these platforms. It underscores the critical role of molecular docking in accelerating drug discovery processes, facilitating computer-aided drug design, and enabling virtual screening methodologies. Furthermore, through compelling case studies, the chapter vividly illustrates the indispensable role of molecular docking in advancing medical development. It underscores its role as a transformative tool driving innovation in medicine development and its significance in shaping the landscape of modern research paradigms.

Acknowledgements

The authors would like to thank Marwadi University Rajkot, Gujarat, India, for providing the necessary assistance to accomplish this study.

Conflict of interest

The authors declare no conflict of interest.

Author details

Neha Mathur[1], Siva Sai Chandragiri[2], Sarita[1], Shristhi Shandily[1],
Krupa Mukeshbhai Santoki[3], Nandini Navinchandra Vadhavana[3], Sejal Shah[3]
and Muktesh Chandra[3]*

1 Amity Institute Pharmacy, Lucknow, Amity University, Noida, Uttar Pradesh, India

2 Department of Pathology, The University of Oklahoma Health Sciences Center, Oklahoma City, OK, USA

3 Faculty of Engineering and Technology, Department of Bioinformatics, Marwadi University, Rajkot, Gujarat, India

*Address all correspondence to: mchandra33@gmail.com;
muktesh.chandra@marwadieducation.edu.in

References

[1] Jorgensen WL. The many roles of computation in drug discovery. Science. 2004;**303**(5665):1813-1818

[2] Bajorath J. Integration of virtual and high-throughput screening. Nature Reviews. Drug Discovery. 2002;**1**(11):882-894

[3] Walters WP, Stahl MT, Murcko MA. Virtual screening—An overview. Drug Discovery Today. 1998;**3**:160-178

[4] Langer T, Hoffmann RD. Virtual screening: An effective tool for lead structure discovery? Current Pharmaceutical Design. 2001;7(7):509-527

[5] Kitchen DB, Decornez H, Furr JR, Bajorath J. Docking and scoring in virtual screening for drug discovery: Methods and applications. Nature Reviews. Drug Discovery. 2004;**3**(11):935-949

[6] Gohlke H, Klebe G. Approaches to the description and prediction of the binding affinity of small-molecule ligands to macromolecular receptors. Angewandte Chemie (International Ed. in English). 2002;**41**(15):2644-2676

[7] Moitessier N, Englebienne P, Lee D, Lawandi J, Corbeil CR. Towards the development of universal, fast and highly accurate docking/scoring methods: A long way to go. British Journal of Pharmacology. 2008;**153**(Suppl. 1):S7-S26

[8] Shoichet BK, McGovern SL, Wei B, Irwin JJ. Hits, leads and artifacts from virtual and high throughput screening. In: Molecular Informatics: Confronting Complexity. 2002

[9] Bailey D, Brown D. High-throughput chemistry and structure-based design: Survival of the smartest. Drug Discovery Today. 2001;**6**(2):57-59

[10] Kuntz ID, Blaney JM, Oatley SJ, Langridge R, Ferrin TE. A geometric approach to macromolecule-ligand interactions. Journal of Molecular Biology. 1982;**161**(2):269-288

[11] DesJarlais RL, Sheridan RP, Dixon JS, Kuntz ID, Venkataraghavan R. Docking flexible ligands to macromolecular receptors by molecular shape. Journal of Medicinal Chemistry. 1986;**29**:2149-2153

[12] Levinthal C, Wodak SJ, Kahn P, Dadivanian AK. Hemoglobin interaction in sickle cell fibers. I: Theoretical approaches to the molecular contacts. Proceedings of the National Academy of Sciences USA. 1975;**72**:1330-1334

[13] Goodsell DS, Olson AJ. Automated docking of substrates to proteins by simulated annealing. Proteins. 1990;**8**:195-202

[14] Salemme FR. An hypothetical structure for an intermolecular electron transfer complex of cytochromes c and b5. Journal of Molecular Biology. 1976;**102**:563-568

[15] Wodak SJ, Janin J. Computer analysis of protein-protein interaction. Journal of Molecular Biology. 1978;**124**:323-342

[16] Kuhl FS, Crippen GM, Friesen DK. A combinatorial algorithm for calculating ligand binding. Journal of Computational Chemistry. 1984;**5**:24-34

[17] DesJarlais RL, Sheridan RP, Seibel GL, Dixon JS, Kuntz ID, Venkataraghavan R. Using shape complementarity as an initial screen in designing ligands for a receptor binding site of known three-dimensional structure. Journal of Medicinal Chemistry. 1988;**31**:722-729

[18] Warwicker J. Investigating protein-protein interaction surfaces using a reduced stereochemical and electrostatic model. Journal of Molecular Biology. 1989;**206**:381-395

[19] Jiang F, Kim SH. "Soft docking": Matching of molecular surface cubes. Journal of Molecular Biology. 1991;**219**:79-102

[20] Meng XY, Zhang HX, Mezei M, Cui M. Molecular docking: A powerful approach for structure-based drug discovery. Current Computer-Aided Drug Design. 2011;**7**:146-157

[21] Amaro RE, Baudry J, Chodera J, Demir Ö, McCammon JA, Miao Y, et al. Ensemble docking in drug discovery. Biophysical Journal. 2018;**114**:2271-2278

[22] Abagyan R, Totrov M. High-throughput docking for lead generation. Current Opinion in Chemical Biology. 2001;**5**:375-382

[23] Carlson HA. Protein flexibility and drug design: How to hit a moving target. Current Opinion in Chemical Biology. 2002;**6**:447-452

[24] Greer J, Bush BL. Macromolecular shape and surface maps by solvent exclusion. Proceedings of the National Academy of Sciences USA. 1978;**75**:303-307

[25] Salmaso V, Moro S. Bridging molecular docking to molecular dynamics in exploring ligand-protein recognition process: An overview. Frontiers in Pharmacology. 2018;**9**:923

[26] De Vivo M, Masetti M, Bottegoni G, Cavalli A. Role of molecular dynamics and related methods in drug discovery. Journal of Medicinal Chemistry. 2016;**59**:4035-4061

[27] Alonso H, Bliznyuk AA, Gready JE. Combining docking and molecular dynamic simulations in drug design. Medicinal Research Reviews. 2006;**26**:531-568

[28] Bard J, Ercolani K, Svenson K, Olland A, Somers W. Automated systems for protein crystallization. Methods. 2004;**34**:329-347

[29] Gavira JA. Current trends in protein crystallization. Archives of Biochemistry and Biophysics. 2016;**602**:3-11

[30] Dauter Z, Wlodawer A. Progress in protein crystallography. Protein and Peptide Letters. 2016;**23**:201-210

[31] Grimes JM, Hall DR, Ashton AW, Evans G, Owen RL, Wagner A, et al. Where is crystallography going? Acta Crystallographica Section D. 2018;**74**:152-166

[32] Durrant JD, McCammon JA. Molecular dynamics simulations and drug discovery. BMC Biology. 2011;**9**:71

[33] Karplus M, McCammon JA. Molecular dynamics simulations of biomolecules. Nature Structural Biology. 2002;**9**:646-652

[34] Rastelli G, Degliesposti G, Del Rio A, Sgobba M. Binding estimation after refinement, a new automated procedure for the refinement and rescoring of docked ligands in virtual screening. Chemical Biology & Drug Design. 2009;**73**:283-286

[35] Hou T, Wang J, Li Y, Wang W. Assessing the performance of the molecular mechanics/Poisson Boltzmann surface area and molecular mechanics/generalized born surface area methods. II. The accuracy of ranking poses generated from docking. Journal of Computational Chemistry. 2011;**32**:866-877

[36] Genheden S, Ryde U. The MM/PBSA and MM/GBSA methods to estimate ligand-binding affinities. Expert Opinion on Drug Discovery. 2015;**10**:449-461

[37] Pu C, Yan G, Shi J, Li R. Assessing the performance of docking scoring function, FEP, MM-GBSA, and QM/MM-GBSA approaches on a series of PLK1 inhibitors. Medchemcomm. 2017;**8**:1452-1458

[38] Rastelli G, Pinzi L. Refinement and rescoring of virtual screening results. Frontiers in Chemistry. 2019;**7**:498

[39] Gschwend DA, Good AC, Kuntz ID. Molecular docking towards drug discovery. Journal of Molecular Recognition. 1996;**9**:175-186

[40] De Vivo M, Cavalli A. Recent advances in dynamic docking for drug discovery. Wiley Interdisciplinary Reviews: Computational Molecular Science. 2017;**7**:e1320

[41] Shoichet BK, McGovern SL, Wei B, Irwin JJ. Lead discovery using molecular docking. Current Opinion in Chemical Biology. 2002;**6**:439-446

[42] Ferreira LG, Dos Santos RN, Oliva G, Andricopulo AD. Molecular docking and structure-based drug design strategies. Molecules. 2015;**20**:13384-13421

[43] Kinnings SL, Liu N, Buchmeier N, Tonge PJ, Xie L, Bourne PE. Drug discovery using chemical systems biology: Repositioning the safe medicine Comtan to treat multi-drug and extensively drug resistant tuberculosis. PLoS Computational Biology. 2009;**5**:e1000423

[44] Pinzi L, Lherbet C, Baltas M, Pellati F, Rastelli G. In silico repositioning of cannabigerol as a novel inhibitor of the enoyl acyl carrier protein (ACP) reductase. Molecules. 2019;**24**:2567

[45] Anighoro A, Pinzi L, Marverti G, Bajorath J, Rastelli G. Heat shock protein 90 and serine/threonine kinase B-Raf inhibitors have overlapping chemical space. RSC Advances. 2017;**7**:31069-31074

[46] Li H, Gao Z, Kang L, Zhang H, Yang K, Yu K, et al. TarFisDock: A web server for identifying drug targets with docking approach. Nucleic Acids Research. 2006;**34**:W219-W224

[47] Lyu J, Wang S, Balius TE, Singh I, Levit A, Moroz YS, et al. Ultra-large library docking for discovering new chemotypes. Nature. 2019;**566**:224-229

[48] Graziani D, Caligari S, Callegari E, De Toma C, Longhi M, Frigerio F, et al. Evaluation of amides, carbamates, sulfonamides, and ureas of 4-Prop-2-ynylidenecycloalkylamine as potent, selective, and bioavailable negative allosteric modulators of metabotropic glutamate receptor 5. Journal of Medicinal Chemistry. 2019;**62**:1246-1273

[49] Ramsay RR, Popovic-Nikolic MR, Nikolic K, Uliassi E, Bolognesi ML. A perspective on multi-target drug discovery and design for complex diseases. Clinical and Translational Medicine. 2018;**7**:3

[50] Lee A, Lee K, Kim D. Using reverse docking for target identification and its applications for drug discovery. Expert Opinion on Drug Discovery. 2016;**11**:707-715

[51] Anighoro A, Bajorath J, Rastelli G. Polypharmacology: Challenges and opportunities in drug discovery. Journal of Medicinal Chemistry. 2014;**57**:7874-7887

[52] Morris GM, Lim-Wilby M. Molecular Modeling of Proteins. In: Kukol A, editor. Totowa, NJ: Humana Press; 2008. pp. 365-382. DOI: 10.1007/978-1-59745-177-2_19

[53] Agu PC, Afiukwa CA, Orji OU, Ezeh EM, Ofoke IH, Ogbu CO, et al. Molecular docking as a tool for the discovery of molecular targets of nutraceuticals in diseases management. Scientific Reports. 2023;**13**(1):13398. DOI: 10.1038/s41598-023-40160-2

[54] Koshland DE. Correlation of structure and function in enzyme action. Science. 1963;**142**(3599):1533-1541. Available from: http://www.ncbi.nlm.nih.gov/pubmed/14075684

[55] Szabo A, Ostlund N S. Modern Quantum Chemistry: Introduction to Advanced Electronic Structure Theory. Mineola, New York, United States: Dover Publications; 1989

[56] Trott O, Olson AJ. AutoDock Vina: Improving the speed and accuracy of docking with a new scoring function, efficient optimization, and multithreading. Journal of Computational Chemistry. Cambridge, Massachusetts, United States. 2010;**31**(2):455-461

[57] Alhajj M, Alhajj R. Clinical trials. In: Encyclopedia of Bioinformatics and Computational Biology. Academic Press; 2019. pp. 550-556. DOI: 10.1016/B978-0-12-809633-8.20451-1

[58] Kelley Lawrence A, Stefans Mezulis Christopher M, Yates Mark N, Wass Michael JE Sternberg. The Phyre2 web portal for protein modeling, prediction and analysis. Nature Protocols 2015;**10**(6):845-858

[59] Altaf A, El Amraoui A, Delmotte F, Lecoutre C. Applications of artificial intelligence in Cross docking: A systematic literature review. Journal of Computer Information Systems. 2023;**63**(5):1280-1300

[60] Vandavasi BNJ, Arunachalam U, Vittal DP, Raju R, Narayanaswamy V, Arumugam V, et al. AI-ML-enabled electromagnetic homing guidance system for scientific autonomous underwater vehicles. Marine Technology Society Journal. 2023;**57**(1):7-15

[61] Clyde A, Liu X, Brettin T, Yoo H, Partin A, Babuji Y, et al. AI-accelerated protein-ligand docking for SARS-CoV-2 is 100-fold faster with no significant change in detection. Scientific Reports. 2023;**13**(1):2105

[62] He L, Rondao D, Aouf N. A novel mechanism for orbital AI-based autonomous refuelling. In: AIAA SCITECH 2023 Forum. Reston, Virginia: American Institute of Aeronautics and Astronautics; 2023

[63] Chavez-Galaviz J, Mahmoudian N. Efficient underwater docking detection using knowledge distillation and artificial image generation. In: 2022 IEEE/OES Autonomous Underwater Vehicles Symposium (AUV). Singapore: IEEE Symposium; 2022. pp. 1-7

[64] Adelusi TI, Oyedele A-QK, Boyenle ID, Ogunlana AT, Adeyemi RO, Ukachi CD, et al. Molecular modeling in drug discovery. Informatics in Medicine Unlocked. 2022;**29**:100880. Available from: https://www.sciencedirect.com/science/article/pii/S235291482200034X

[65] Andrusier N, Mashiach E, Nussinov R, Wolfson HJ. Principles of flexible protein-protein docking. Proteins. 2008;**73**(2):271-289. Available from: http://www.ncbi.nlm.nih.gov/pubmed/18655061

[66] Gaudreault F, Baardsnes J, Martynova Y, Dachon A, Hogues H, Corbeil CR, et al. Exploring rigid-backbone protein docking in biologics discovery: A test using the DARPin scaffold. Frontiers in Molecular Biosciences. 2023;**10**. Available

from: https://www.frontiersin.org/articles/10.3389/fmolb.2023.1253689/full

[67] Alogheli H, Olanders G, Schaal W, Brandt P, Karlén A. Docking of macrocycles: Comparing rigid and flexible docking in glide. Journal of Chemical Information and Modeling. 2017;**57**(2):190-202. DOI: 10.1021/acs.jcim.6b00443

[68] Prieto-Martínez FD, López-López E, Juárez-Mercado KE, Medina-Franco JL. Chapter 2-Computational Drug Design Methods—Current and Future Perspectives. In: Kunal R, editor. Silico Drug Design. Academic Press. 2019:19-44. ISBN 9780128161258

[69] Vina A, Oleg T, Olson Arthur J. Improving the speed and accuracy of docking with a new scoring function, efficient optimization, and multithreading. Journal of Computational Chemistry. 2010;**31**(2):455-461

[70] Friesner RA, Banks JL, Murphy RB, Halgren TA, Klicic JJ, Mainz DT, et al. Glide: A new approach for rapid, accurate docking and scoring. 1. Method and assessment of docking accuracy. Journal of Medicinal Chemistry. 2004;**47**(7):1739-1749

[71] Abagyan R, Totrov M, Kuznetsov D. ICM—A new method for protein modeling and design: Applications to docking and structure prediction from the distorted native conformation. Journal of Computational Chemistry. 1994;**15**(5):488-506

[72] Jones G, Willett P, Glen RC, Leach AR, Taylor R. Development and validation of a genetic algorithm for flexible docking. Journal of Molecular Biology. 1997;**267**(3):727-748

[73] Rarey M, Kramer B, Lengauer T, Klebe G. A fast flexible docking method using an incremental construction algorithm. Journal of Molecular Biology. 1996;**261**(3):470-489

[74] Ewing TJ, Makino S, Skillman AG, Kuntz ID. DOCK 4.0: Search strategies for automated molecular docking of flexible molecule databases. Journal of Computer-Aided Molecular Design. 2001;**15**(5):411-428

[75] Allen WJ, Balius TE, Mukherjee S, Brozell SR, Moustakas DT, Lang PT, et al. DOCK 6: Impact of new features and current docking performance. Journal of Computational Chemistry. 2015;**36**(15):1132-1156

[76] Kubinyi H. Success stories of computer-aided design. In: Computer Applications in Pharmaceutical Research and Development. 2006. pp. 377-424. DOI: 10.1002/0470037237.ch16

[77] Irwin JJ, Shoichet BK. ZINC--A free database of commercially available compounds for virtual screening. Journal of Chemical Information and Modeling. 2005;**45**(1):177-182. DOI: 10.1021/ci049714+

[78] Jain AN. Surflex-dock 2.1: Robust performance from ligand energetic modeling, ring flexibility, and knowledge-based search. Journal of Computer-Aided Molecular Design. 2007;**21**:281-306

[79] Huey R, Morris GM, Olson AJ, Goodsell DS. A semiempirical free energy force field with charge-based desolvation. Journal of Computational Chemistry. 2007;**28**(6):1145-1152

[80] Warren GL, Andrews CW, Capelli AM, Clarke B, LaLonde J, Lambert MH, et al. A critical assessment of docking programs and scoring functions. Journal of Medicinal Chemistry. 2006;**49**(20):5912-5931

[81] Yang W, Lai L. Computational design of ligand-binding proteins. Current Opinion in Structural Biology. 2017;**45**:67-73

[82] Verma S, Patel CN, Chandra M. Identification of novel inhibitors of SARS-CoV-2 main protease (Mpro) from Withania sp. by molecular docking and molecular dynamics simulation. Journal of Computational Chemistry. 2021;**42**(26):1861-1872

[83] Chaudhary KK, Mishra N. A review on molecular docking: Novel tool for drug discovery. Database. 2016;**3**(4):1029

[84] Lyskov S, Gray JJ. The RosettaDock server for local protein–protein docking. Nucleic Acids Research. 2008;**36**(suppl_2):W233-W238

[85] Torchala M, Moal IH, Chaleil RAG, Fernandez-Recio J, Bates PA. SwarmDock: A server for flexible protein–protein docking. Bioinformatics. 2013;**29**(6):807-809

[86] Cross SSJ. Improved FlexX docking using FlexS-determined base fragment placement. Journal of Chemical Information and Modeling. 2005;**45**(4):993-1001

[87] Schneidman-Duhovny D, Inbar Y, Nussinov R, Wolfson HJ. PatchDock and SymmDock: Servers for rigid and symmetric docking. Nucleic Acids Research. 2005;**33**(suppl_2):W363-W367

[88] Thomsen R. Christensen MH. MolDock: A new technique for high-accuracy molecular docking. Journal of Medicinal Chemistry, 2006;**49**(11):3315-3321. DOI: 10.1021/jm051197e

[89] Jain AN. Surflex: Fully automatic flexible molecular docking using a molecular similarity-based search engine. Journal of Medicinal Chemistry. 2003;**46**(4):499-511

[90] Abagyan R, Totrov M. Biased probability Monte Carlo conformational searches and electrostatic calculations for peptides and proteins. Journal of Molecular Biology. 1994;**235**(3):983-1002

[91] Tahir ul Qamar M, Bari A, Adeel MM, et al. Peptide vaccine against chikungunya virus: Immuno-informatics combined with molecular docking approach. Journal of Translational Medicine. 2018;**16**:298. DOI: 10.1186/s12967-018-1672-7

[92] Firdaus MER, Mustopa AZ, Triratna L, Syahputra G, Nurfatwa M. Dissection of capsid protein HPV 52 to rationalize vaccine designs using computational approaches immunoinformatics and molecular docking. Asian Pacific Journal of Cancer Prevention. 2022;**23**(7):2243-2253. DOI: 10.31557/APJCP.2022.23.7.2243

[93] Bazmara S, Shadmani M, Ghasemnejad A, Aghazadeh H, Bagheri KP. In silico rational design of a novel tetra-epitope tetanus vaccine with complete population coverage using developed immunoinformatics and surface epitope mapping approaches. Medical Hypotheses. 2019;**130**:109267

[94] Sasaki-Tanaka R, Nagulapalli Venkata KC, Okamoto H, Moriyama M, Kanda T. Evaluation of potential anti-hepatitis a virus 3C protease inhibitors using molecular docking. International Journal of Molecular Sciences. 2022;**23**(11):6044. DOI: 10.3390/ijms23116044

[95] Arasu MV, Vijayaragavan P, Purushothaman S, Rathi MA, Al-Dhabi NA, Gopalakrishnan VK, et al. Molecular docking of monkeypox (mpox) virus proteinase with FDA approved lead molecules. Journal of Infection and Public Health. 2023;**16**(5):784-791. DOI: 10.1016/j.jiph.2023.03.0

Chapter 3

Global Ligand-Protein Docking Tools: Comparation and Case Study

Vy T.T. Le, Tu H.T. Nguyen and Phuc-Chau Do

Abstract

Molecular docking study, a method used in drug discovery, is used to estimate the interactions between small molecules and macromolecules. Docking can be divided into rigid and flexible docking where local and global docking is the subclass in the flexible approach. Two important criteria in docking are search algorithms and scoring functions. The former assesses the fitness of ligand poses within the protein's binding site, while the latter explores different ligands "conformations until the point at which the least energy convergence is achieved." Three user-friendly global docking programs (AutoDock Vina, MOE, and DOCK6) are utilized to study ligand behaviors against *Enterovirus A71* 3C protease ($3C^{pro}$), which causes hand-foot-mouth disease in children. The results suggested that the DOCK6 gives the fastest output, and all of the ligands correctly bind to the active site of $3C^{pro}$. Rupintrivir is a good candidate for serving as a positive control in all three tools for binding site identification because it shows broad resistance to viruses. In comparison to AutoDock Vina and MOE, DOCK6 exhibits superior conformational space search efficiency and speed across the three docking technologies used in our investigation. AutoDock Vina, however, is typically regarded as being more appropriate for novices.

Keywords: scoring functions, search algorithms, rigid and flexible docking, global docking, AutoDock Vina, MOE, DOCK6, *Enterovirus A71*, 3C protease

1. Introduction

In the fields of molecular biology and pharmacology, studying how ligands interact with proteins is crucial because it plays a key role in comprehending basic biological processes and developing therapeutic treatments. Ligands can be tiny molecules, medications, or signaling molecules. Its function is to interact with proteins to influence a variety of biological processes (e.g., enzymatic activity, signal transmission, and gene expression). Scientists can gain crucial understanding of the fundamental processes of disease etiology and possible paths toward treatment by clarifying the molecular processes that underlie these interactions. Knowing how ligands interact with proteins makes it easier to predict drug toxicity, efficacy, and side effects, as well as to identify adverse reactions. This knowledge aids in optimizing drug candidates for better pharmacokinetic and bioavailability during drug discovery and development. Precision medicine techniques are made possible by the rational creation of innovative treatments that are specifically targeted at molecular targets or certain disease pathways, which is made possible by a thorough understanding of ligand-protein interactions. The investigation of ligand-protein interactions is fundamental to the advancement of pharmaceutical research, propelling breakthroughs in drug discovery and development, and to our comprehension of biology and disease.

Biological systems are incredibly flexible and adaptive; they can respond to a variety of stimuli and maintain equilibrium. The intricate nature of signaling pathways and transcriptional networks is demonstrated by the dynamic regulation of gene expression. Their intricacy is highlighted by the way that many elements interact, such as through protein-protein interactions and metabolic pathways. Understanding emergent behaviors is hampered by breaking down distinct components. Understanding the dynamic nature of biological systems requires the application of integrated viewpoints. Due to their transient and reversible character, as well as their dynamic conformational changes, protein-ligand interactions present difficulties in these kinds of systems. Interactions are further complicated by varying cellular architecture and subcellular localization patterns. Temperature, pH, and posttranslational changes are among the factors that make it difficult to measure interaction specificity precisely.

Drug development problems require multidisciplinary approaches that combine computational modeling and experimental methods, such as structural biology. In order to assist identify leads against macromolecules, virtual screening and molecular docking have become effective methods for high-throughput screening Molecular docking is famous for identifying novel lead compounds rapidly against macromolecules since 1980s due to the significant growth in computing power and abundant availability of molecular structures [1]. These methods provide insights into molecular mechanisms and binding affinities by modeling interactions between proteins and ligands. To aid in the search for new drugs, docking studies forecast ligand selectivity and off-target effects. Understanding protein-ligand interactions in biological systems is improved by combining experimental techniques with docking simulations [1].

To perform docking, two approaches can be applied, namely, rigid and flexible docking. A report carried out by Xuan-Yu Meng and his colleagues defined a rigid docking where ligands and proteins maintain fixed conformations during docking process, neglecting the flexibility and dynamic movements of the molecules. Then, Fischer proposed a new approach, which is to measure the geometric fit between their shapes is directly comparable to their affinity, and is known as the lock-and-key theory [2]. Specifically, in the simplest rigid-body systems, to fit the protein's binding site, ligands undergo the search in a six-dimensional space including rotational and translational changes *via* multiple interactions (e.g., van der Waals forces, coulombic interactions, and hydrogen bonds). As a result, docking poses are generated as the output, which will be ranked according to their scores, and the most favorable ones will be selected, which might imply the active site of that protein. However, the rigid body docking does not satisfy the experimental reality that molecules, especially proteins, are crystallized separately. The three-dimensional structure of protein primarily determined its function. This is defined by intricate macromolecules composed of linear chains of amino acids, is highly dynamic, and constantly changes conformations, which can significantly affect the overall shape complementary and interactions with ligands [3]. Moreover, assessment of accuracy in docking is preferable in bound complexed rather than unbound ones, indicating that treating proteins as entire rigid body is not certified [1]. The induced fit approach was first introduced by Koshland et al. in 1963. It suggested that proteins and ligands be flexible during docking to accommodate the dynamic nature of the system [4]. Partial or fully flexible methods, where local changes or entire protein conformations are allowed to vary, are included in flexible docking. While fully flexible docking incorporates side chain and backbone changes, partial flexibility concentrates on particular regions of the molecular structure. Through the investigation of various conformational changes, these techniques bolster the binding mode's accuracy as well as predict binding affinity efficiently. Large conformational spaces can be efficiently explored with the help of technological advancements, such as graphics processing units (GPUs), which increase computing efficiency for extensive searches [1].

2. Global docking for protein-ligand system

Global docking or blind docking refers to a method where a ligand (small molecules) will be docked to a whole surface of a target receptor (typically a protein) without knowing prior to the receptor's binding site. To be more specific, the process searches for the entire conformational space of both ligand and receptor to determine the most favorable interaction and binding poses (**Figure 1**). While local docking focuses on flexibility at the specific site of docking, global docking allows all the rigid, flexibility, and orientation of both molecules, which might require more resource requirements and running time.

Good global docking results require careful consideration of numerous crucial parameters. The precision of the protein structure employed for docking is critical. Docking accuracy improves when both the protein and the ligand are properly prepared, including the removal of water molecules and the adjustment of protonation states. Furthermore, it is critical to select appropriate scoring systems that accurately measure binding affinity and to incorporate flexibility in both the protein and the ligand during simulation. Docking results can be validated against experimental data to ensure trustworthiness, while efficient sampling techniques and taking into account solvent effects enhance prediction accuracy. Recently, there have been more

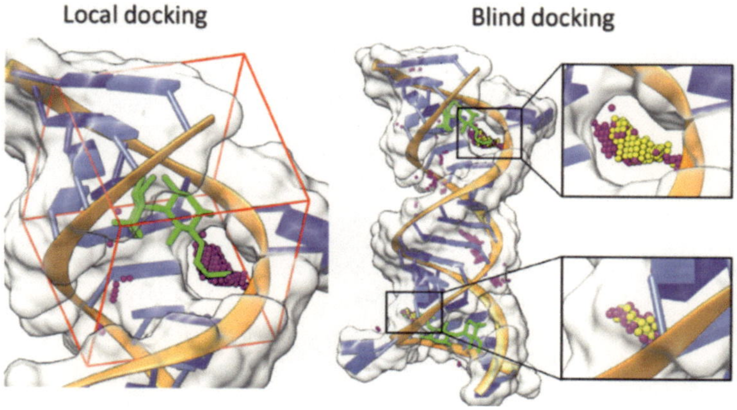

Figure 1.
Local docking and global docking of an aminoglycoside antibiotic, gentamicin (green), and the 16SrRNA a site of bacterial ribosome. Pink and yellow residues are the binding sites estimated by RLDOCK [5].

than 60 docking programs and servers appearing to be available for academic or commercial purposes [6–8]. The differences among these docking tools are a scoring function and a search algorithm in estimating the binding modes and affinities between proteins and ligands (**Table 1**). The capability of sufficiently sampling the degrees of

No	Tools	Year	Scoring functions	Search algorithms	Lisence
1	ICM [9]	1994	Internal Coordinate Mechanics scoring	Monte Carlo, Recursive Belief-based Learning	Academic free
2	Flex X [10]	1996	Ludi score, Chem score, FlexX score	Incremental Construction, Branch and Bound	Commercial
3	Gold [11]	1997	Gold Score, Chem Score, Astex Statistical Potential, ChemPiecewise Linear Potential	Genetic Algorithm, Monte Carlo	Commercial
4	DOCK6 [12–16]	2003	Contact score, Grid-based score, DOCK 3.5 score, Continuous score, Zou Generalized-Born/ surface area, Hawkins Generalized-Born/ surface area, AMBER score, Footprint score, Multigrid score, Pharmacophore Matching Similarity, Internal Energy Simple Score, Solvent Accessible Surface Area score, Descriptor score	Anchor-and-grow, De novo, Genetic Algorithm, sphere-matching	Academic free
5	Fred (Fast Rigid Exhaustive Docking) [17]	2003	Screenscore, Chemgauss, Shapegauss, ChemScore and Piecewise Linear Potential	Shape-based (Gaussian)	Commercial
6	Surflex [18]	2003	Surflex score	Broyden-Fletcher-Goldfarb-Shanno	Commercial
7	Gemdock [19]	2004	Simple empirical scoring function and a pharmacophore-based scoring function	Evolutionary algorithm	Free
8	AutoDock Vina [20, 21]	2010	Standard Vina scoring	Monte Carlo-Broyden-Fletcher-Goldfarb-Shanno	Academic free
9	Symmref [22]	2011	Energy scoring function	Symmref algorithm, Monte Carlo-Broyden-Fletcher-Goldfarb-Shanno	Free
10	MOE [23]	2012	Systematic and Stochastic sampling	Atom-sphere exclusion score, Affinity ΔG score, Alpha Hydrogen Bond score, London ΔG score and Generalized Born-Volume Integral/ Weighted Surface Area ΔG score	Academic free

No	Tools	Year	Scoring functions	Search algorithms	Lisence
11	LeDock [24]	2013	Chemscore, Piecewise Linear Potential	Simulated annealing and genetic algorithm	Free
12	PyRx [25]	2015	Random Forest-based scoring function for Virtual Screening	Monte Carlo-Based Search, Lamarckian Genetic Algorithm, Iterated Local Search, Local Optimization	Free

Table 1.
Selected global docking tools.

freedom (rotation and translation) within the complex system to ensure the actual binding modes are counted has proven to be an effective characteristic of basic search algorithms. Meanwhile, scoring function's role is to predict the binding free energies (Δ G bind) within a specific range, followed by the evaluation of the true binding modes and finally ranking poses based on predicted binding affinities.

2.1 Searching algorithm

Effective methods are essential for precise global docking, wherein the root mean square deviation (RMSD) is employed to measure ligand position in relation to experimental data. Due to computational limitations (considering just six degrees of freedom: rotation, translation, and conformational change), rigid body approximation was initially popular. However, it has constraints in describing particular interactions between ligands and binding sites. Certain programs maximize computing resources, improve interaction realism, and provide complete ligand variation while restricting the flexibility of proteins. Conformational alterations in ligands or receptors are referred to as flexibility in docking. The three types of flexible ligand docking methodologies are simulation, stochastic, and systematic searches.

The aim of the systematic search methods is to systematically cover the entire search area of possible ligand's configurations. They are subsequently divided into three subcategories, namely, exhaustive, fragmentation, and conformational ensemble methods. The first method, exhaustive search, involves the rotation at 360° of all possible rotatable bonds in ligand at predefined intervals, but some challenges arise with combinatorial explosion, restricting its practicality in exclusively small and flexible ligands [6]. Meanwhile, fragmentation approach splits the ligands into modular parts, and one of its fragments is placed immovably in the receptor's binding site, followed by adding other flexible parts to complete the ligand structure. Regarding conformational ensemble approach, despite minimizing the computational costs because pre-formed ligand conformations are used, generating such conformational ensembles requires additional tools and bioactive structures may sometimes be neglected.

In the context of stochastic algorithms, random alterations based on predefined probability functions are either accepted or rejected to seek ligand binding orientations and conformations. Hence, a global energy minimum has risen due to a variety of conformational space of ligands, as well as energy landscapes discovered. However, as a large search space is examined, this results in computational extensive. There are four basic categories: Genetic Algorithm (GA), Monte Carlo (MC), Ant Colony Optimization (ACO), and Tabu Search (TS). The first group-MC method was primarily applied in the minimization step in molecular dynamic (MD) simulations (e.g., GROMACS [9]). Its principle lies in applying the Boltzmann probability function to calculate the chance of accepting random fluctuations in thermodynamically accessible states, which decreases the chances of being trapped in local minima [10]. Secondly, GAs are derived from evolutionary programming algorithms (employ evolution-driven computational models) designed to determine the precise conformation to the global energy minimum. Meanwhile, TS method is an iterative optimization process and a variation of GA method [1]. It functions as a meta-heuristic method that can switch from one solution to another while keeping previously visited solutions in memory. However, TS separates itself from local search strategies by integrating a memory structure (Tabu list) to prevent the return of already guessed solutions and to encourage further exploration of new regions [11]. The principle of the last ACO type aims to apply pheromone trails for marking the low-energy ligand structures, which are altered in further repetitions to produce apparently low-energy conformations [12].

While systematic search focuses on thoroughly exploring the whole conformational space in a methodical manner and the objective of stochastic search is to select the optimal ligand conformations based on a predetermined probability function, simulation search involves solving Newton's equations of motion. MD and pure energy minimization methods stand out as the two main types of this approach. The former considers forces and motions in the molecular behaviors over the time, but it is limited in locating rugged hypersurfaces and productively sampling different ligand structures within an appropriate simulation period. Meanwhile, the latter requires the most robust system possibly by reducing the potential energy of the system. It comprises diverse techniques such as direct searches, gradient methods, or conjugate-gradient methods (e.g., Fletcher-Reeves) and is often used complementarily with other docking algorithms [13–15].

Several docked structures with favorable surface complementarity are built on the topic of rigid body approximation and then reorganized in relation to free energy. The fast Fourier transform (FFT) correlation method explores systems docked structures *via* electrostatic interactions or a combination of electrostatic and solvation terms [6]. Other features, such as atomic interaction properties, relative shape, dissociation energies, polar Fourier interactions, and advanced software/hardware technologies to accelerate outputs, are incorporated in some FFT-based methods [16, 17]. Another algorithm utilized in rigid body docking, which aims to determine complementarily molecular shapes using such surface patches or 3D Zernike descriptors and thus aid better alignment of local geometric feature, is known as the geometric hashing (GH) method [18]. Likewise, shape matching (SM) algorithms evaluate geometric overlap in identifying optimal matching between ligands and receptor's binding site [19]. DOCK tool – a pioneer in employing both SM and GH algorithms to recognize potential binding sites – sphere centers [20].

2.2 Scoring function

With the advance in computational hardware, the primary challenges have shifted from search-related algorithms to those involving scoring. Specifically, the challenges in Δ G bind's prediction should be considered, such as complex physical interactions – entropy and enthalpy (e.g., interacting bonds, solvent molecules, entropy loss due to rotation and translation of ligands). This leads to the idea of being simplified and assumptive in scoring functions to decrease the complexity of the systems, as well as balancing the speed and accuracy are crucial to avoid errors. Ideally, an optimal search algorithm comes along with the best scoring function, which is contrary to current docking tools depending on the specific characteristics of receptor's binding sites and studied ligands, so making the method become impractical [10]. In the realm of scoring functions, it is categorized into three basic groups, namely, physics-based, empirical, and knowledge-based functions to forecast how strong the interactions between proteins and ligands are (**Table 2**). Each class contributes to advancing

Aspects	Physical-based	Empirical	Knowledge-based
Principle	Enthalpy contribution is considered along with the sum of bonded and non-bonded interactions within the complex.	The addition of different energetic terms ranging from hydrogen bonds to hydrophobic effects is then multiplied with a coefficient derived from linear regression analysis.	The distance-dependent pairwise potentials are transformed from the atom pairs using the Boltzmann law after obtaining the occurrence frequencies of the protein-ligand complexes' structural information in a database
Advantages	Evaluation the binding poses quickly, suitable for high-throughput docking. Alignment with modern force fields Obtaining from fundamental physics principles Computational cost-effective	Simple energy term treatment makes faster computation Preferable for virtual screening	Affordable for large scale database Robust to training set diversity Effective differing the binding modes. Extension to many-body interactions More variable in ligand flexibility
Challenges	Restriction in approximations Ignoring the long-range binding effects, solvent effects and entropic terms makes inaccurate for complex and real biological systems. Rigid body assumptions are challenges in ligand and protein variations	Not consider the involvement of solvents explicitly Multiple energy terms are double-counting issue. Limited description of flexibility Reliant on training set quality causes lack of generality	Reference state determination Issues relating to potential energy accuracy. Less sensitive to solvent effects Entropic energy makes difficulty in accurate calculation. Sensitive to ligand positions
Recent advances	Refinements in electrostatic interactions, ligand desolvation energies. Post-scoring docking applying the treatment of solvent molecules as continuum dielectric medium to accelerate force field models.	Introduction of continuous, non-linear terms and ligand poses' variation in functions leads to the Hammerhead and Surflex functions. Machine-learning approaches in some empirical scoring functions	Iterative methods for improving accuracy in ITScore, including solvation effects and configurational entropy. Quantitative structure-activity relationships (QSAR)-Machine-learning (ML) approach based on distance-dependent atom pair occurrences.
Examples	Goldscore, AutoDock, Generalized Born/ surface area model	Chemscore, Glidescore, ChemPLP	Drugscore, RF-score, ITscore.

Table 2.
Comparison among three basic scoring function [1, 10, 21–23].

our understanding of molecular docking and holds implications for various applications from lead optimization to virtual screening in drug discovery.

In addition to the basic scoring functions mentioned above, many improved scoring approaches have been evaluated, but they are limited to a specific task [1]. Thus, scientists have raised up the idea of combining multiple scoring functions to tackle the errors from individual function, which is termed the consensus scoring method. In order words, best-docked pose of each compound will be reassessed using diverse scoring functions and the compounds that frequently ranked top scored across all functions are determined as potential candidates for further assays [10]. Take X-SCORE as an example of consensus scoring technique that combines various scoring functions or algorithms such as ChemScore, FlexX, DOCK-like, and GOLD-like [24]. Despite being concerned about error amplification, consensus scoring function still proves its capability in ranking effectiveness or false positives reduction. Another approach that is developed these recent years is to apply ML – the random forests, and the convolutional neutral network – in scoring function, principally introducing QSAR analysis to assess the interaction between proteins and ligands [1]. This method involves multiple properties of the complex such as atom pairs, geometric factors, and ligand-based characteristics to build models for evaluating the binding scores. The process involves two stages. Initially, ML automatically learns the known structure and binding information from a training dataset, which is similar to empirical scoring approach. Then, ML will be trained and builds its own mathematical formulas to rank and screen with better accuracy than other basic scoring approaches. Some relevant ML scoring functions are random forest-based score, and support vector machine score [25].

Above algorithms are generally applied for flexible ligand docking, which could be employed for flexible receptor docking, but obtaining the results at a reasonable timeframe is not yet addressed due to an increased in dimensionality and hence computationally intensive for a larger search space. One solution is to use intricated algorithms combined with simulations to effectively locate the expanded search space, but this still accounts for too much computational resources. Other well-known approaches, including rotamer libraries, soft-receptor modeling, and ensembles of protein conformations have been developed with the purpose of simplifying the system's approximation, accompanied by a reduction in computational costs. The transition to using different global docking tools on a given system necessitates a thorough examination of each tool's performance characteristics. This evaluation ought to consider a variety of factors such as the accuracy of anticipated binding poses and binding affinities, the productivity of computing resources employed, and the algorithm's resistance to fluctuations in input parameters. Furthermore, assessing each tool's virtues and drawbacks reveals important information about its underlying algorithms and approaches. Such extensive comparison studies allow researchers to make educated judgments about which docking tool is most suited to their individual research needs, enhancing the accuracy and usefulness of docking results to research objectives.

In this book chapter, we carried out a case study on the 3C protease ($3C^{pro}$) of *Enterovirus* (*EV*) – *A71* that causes hand-foot-mouth disease (HFMD) in mostly children worldwide. A series of ligands, which have been assessed for their antiviral properties against different mutated *EV-A71* $3C^{pro}$, were tested using three common docking tools AutoDock Vina [26, 27], MOE [28], and DOCK6 [20]. Subsequently, their docking poses were assessed to identify the active site based on their proximity to the binding site. Ultimately, a comparative analysis on different criteria (e.g., execution time, setting, process on multiple ligands) was also carried out for further conclusion.

3. Case study in 3Cprotein of Enterovirus A71

HFMD, which is primarily caused by *EV* and *Coxsackievirus* (*CV*), belonging to Picornaviridae family, is a pediatric disease mostly targeting children under the age of five [29]. Reports found out that *EV-A71* displays more severe consequences such as dysfunction in the central nervous system, cardiopulmonary failure, or vesicular eruptions on feet, hands, and oral mucosa than *CV-A16* or other strains [30]. *EV-A71* was initially discovered in 1957 in New Zealand and then isolated from an encephalitis female's face in 1969 in California [30, 31]. Then, in the late twentieth century, many outbreaks of HFMD had been detected in all over the world, and most large expansions of this serious disease occurred in Asia-Pacific areas, which resulted in unattainable severity tracking. Unfortunately, there are limitations in available vaccines against these strains of *EV-A71* and only symptomatic-supportive treatments are available [36]. Thus, these have raised an emergency alarm for thorough studies on different strains with existing inhibitors as well as leveraging for new potential candidates against *EV-A71*.

EV-A71, which attaches to host cells by different receptors, has a single-stranded RNA genome. This facilitates the uncoating and viral replication processes [32]. There are four categorized regions in *EV-A71* genome denoting from P1–P3 in which P1 is responsible for VP1–VP4 while seven non-structural proteins belong to P2 and P3 groups. During the maturation of *EV-A71*, 3C protease ($3C^{pro}$) – one of the P3's subclasses, which is a chymotrypsin-like fold structure, involves in cleaving the viral polyprotein at eight distinct sites [33]. Another role of $3C^{pro}$ is the cleavage of essential eukaryotic initiation factors, which results in preventing host cap-dependent translation [34]. A report carried out by Wang et al. revealed that the binding site of $3C^{pro}$ contains three key amino acid residues (His40, Glu71, and Cys147) which is responsible for cooperatively promoting substrate cleavage [35]. Additional to the catalytic important loop is the $3C^{pro}$ cleavage site involving such several main amino acid residues: Tyr122, Phe124, Glu126, Leu125, Leu127, Ser128, Thr142, His161, Ile162, Gly164, and Phe170 [36]. In our previous study, these two mentioned binding sites, which had been discovered earlier by Wang and Lu' groups, are preferable for most inhibitors [35, 37]. Thus, these binding sites could be considered as the active site of $3C^{pro}$ proteins.

Furthermore, viral mutations are unavoidable, resulting in the formation of new viral variants with changed antigenic features, which may lead to immune evasion and reduced efficiency of current vaccines and treatments, as well as research chemicals. Furthermore, some mutations may give resistance to antiviral medications, reducing their effectiveness in treating viral infections. Thus, understanding how mutations affect the binding of ligands globally is crucial in order to cope with the viral evolution and variations in $3C^{pro}$ sequences.

In our study case, three models of $3C^{pro}$ proteins are selected, including the wild-type strain (PDB ID: 5C1U) and two mutant strains (PDB ID: 5GSW, 3QZQ). Besides, five prospective ligands are considered for docking. Rupintrivir, one of the synthetic drugs used to treat HFMD, has proven to pass the first phase of clinical test successfully. Meanwhile, two additional synthetic compounds – an isopropyl-substituted 4-iminooxazolidin-2-one moiety (FIOMC) and an α -keto amide derivative (8x),

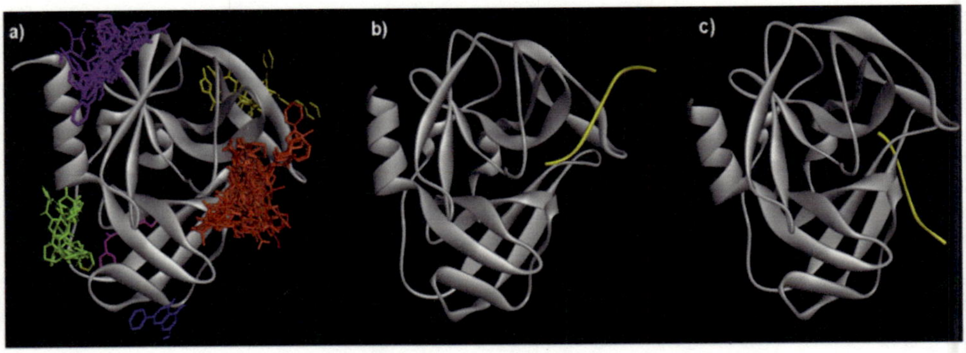

Figure 2.
Binding sites of $3C^{pro}$ protein (PDB ID: 5C1U) with a) six clusters generated when 5C1U docked by AutoDock Vina with 21 control candidates, b) experimental 3SJ9, and c) 3SJK between $3C^{pro}$ (gray color) with its targeted peptide (yellow color) [36].

alongside with two naturally occurring compounds belonging to the flavonoid family (rutin and luteoloside) – are chosen due to their potent inhibitory effects on the $3C^{pro}$ proteins [36]. The active site of $3C^{pro}$ proteins was identified in our prior work and displayed in **Figure 2** below. Hence, ligands residing at the peptide's position are recognized as the promising antiviral substances against $3C^{pro}$ proteins.

3.1 AutoDock Vina

AutoDock Vina is widely used due to its user-friendly interface for molecular docking and virtual screening. In 2009, Dr. Oleg Trott and his team at molecular graphics lab at the Scripps Research Institute, USA, developed this tool from AutoDock 4.0 (AD4.0) [38]. Then, AutoDock Vina 1.2.0 was reported by Jerome et al. with additional support in the treatment of water molecule explicitly and multiple ligands docking (cite). Currently, a newest version of AutoDock Vina 1.2.5 is available at center for computational structural biology website (*https://vina.scripps.edu/*), belonging to Dr. Oleg's lab. Alongside with other docking engines in AutoDock Suite package [38–40], AutoDock Vina recognizably outperforms with AD4.0 for being 100 times faster, as well as significantly reduce the running times *via* exploiting multiple CPUs or CPU cores. Moreover, its accuracy in binding mode predictions is also upgraded by employing efficient search algorithms and assessing fewer scoring functions. In addition, some parameters, that have proven to be well-performed in diverse scenarios, are automatically set. This allows the software to be more accessible and straightforward for either experts or nonexperts in the field.

AutoDock Vina's configuration file has adjustable options that accommodate different system sizes and docking goals. Coordinates and box size are adjusted in global docking scenarios to sufficiently encompass the target region. Docking performance can be adjusted with flexibility using parameters, such as exhaustiveness value (default = 8) and energy range (default = 4 kcal/mol). While more exhaustiveness improves conformational exploration, it also increases computing times. Energy range modification impacts docking diversity and fidelity, as well as posture selection strictness. Rupintrivir's broad antiviral spectrum is shown by the AutoDock Vina results, which show preferential ligand binding to active sites (**Figure 3**).

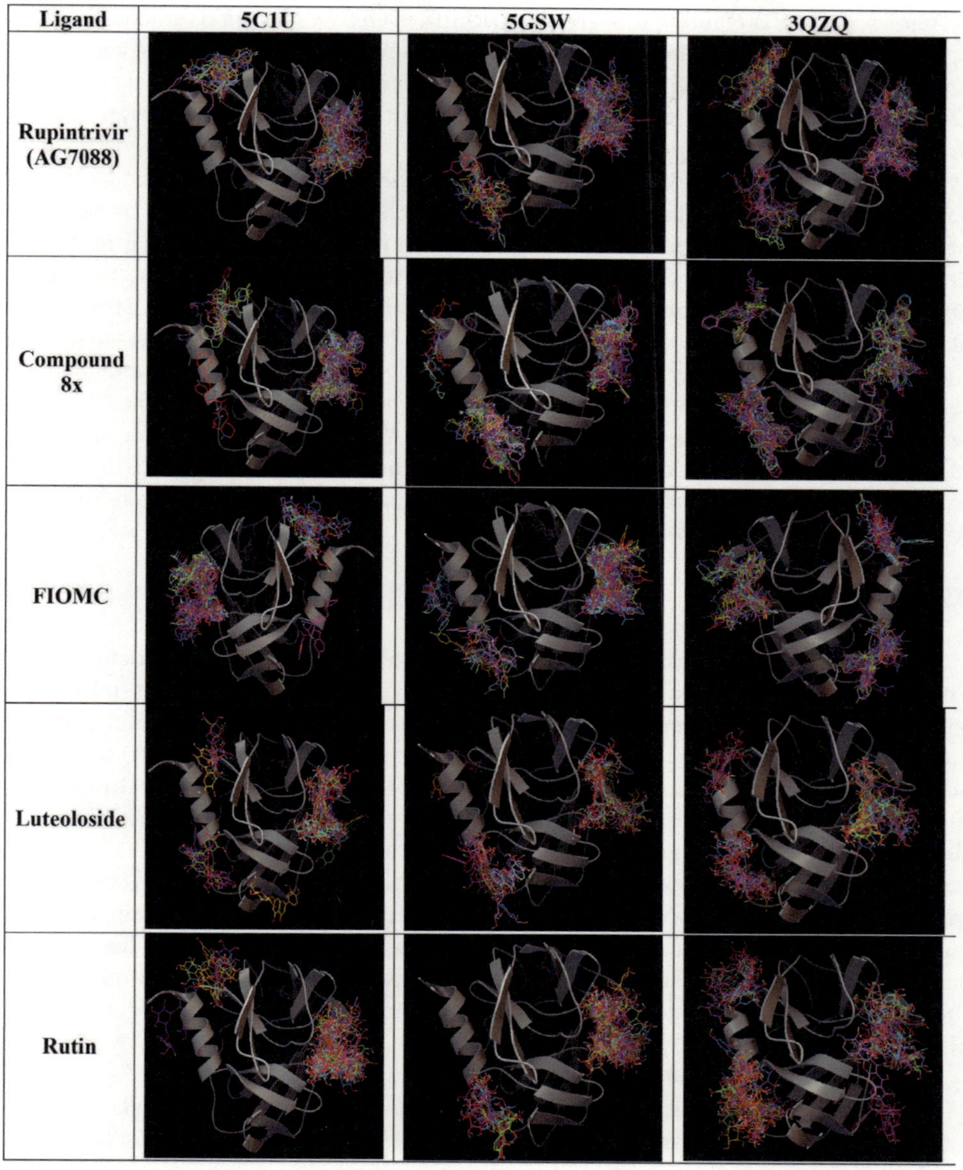

Figure 3.
Binding poses formed by AutoDock Vina between control ligands (line presentation) on three 3C^{pro} models (ribbon presentation).

As mentioned above, scoring functions and search algorithms are the major features in assessing the performance of a docking software. In the latest AutoDock Vina's version, it includes both AD4.2 and Vina scoring functions. This is because Vina lacks some parameters that are present in a physics-based model – AD4.0 such as directional hydrogen bond potentials, electrostatic, and solvation [26]. Additionally, deriving from both knowledge-based and empirical scoring functions, the ligand conformational space and its interactions with receptors are assessed thoroughly. This effectively heightens the system's accuracy. Pertaining to search algorithms, even though various stochastic global optimization approaches have been exploited in the evolution of Vina. Thus, a similar method developed by Abagyan et al. is known as Iterated Local Search global optimizer [27]. Thanks to the ability in using multicore CPUs in Vina, this intensive iteration can be done efficiently in an acceptable timeframe.

However, this is contradicted to what we researched earlier using the exhaustiveness value of 256 that FIOMC displays the best antiviral characteristic across the broad range of virus strains. Besides, the number of binding sites in the wild-type strain is more condensed than in the others. With regard to two mutant proteins, 5GSW is mutated from an asparagine to serine at position 69, while the glutamic acid is mutated to aspartic acid at position 71 in 3QZQ. It is noticeable that the ligands tend to be more dispersed than in the case of 5GSW, which agrees to our previous study [36]. Regarding the average time for one $3C^{pro}$ model with five ligands, it is estimated about 2 hours for a triplicate run with a default exhaustiveness value. This is because AutoDock Vina allows for maximum of 20 modes per single docking performance, so each pair of protein ligand is docked three times with different random seed numbers in order to produce more binding poses for a better evaluation, as well as reproducibility.

3.2 DOCK6

DOCK was initially built by Sheridan et al., implementing geometric matching and distance-based comparisons to find the most favorable binding poses in 1986 [41]. Since then, it has been remarkably evolving with additional essential features, such as Hawkins-Cramer-Truhlar Generalized Born/Surface Area (GB/SA) solvation scoring and Poisson–Boltzmann/Surface Area (PB/SA) solvation scoring through OpenEye's Zap Library. This leads to the release of the newest version of DOCK6 developed by Therese Lang et al. and his collaborators in 2009 [20, 42–45]. This new version has more functions, such as receptor flexibility consideration and conjugate-gradient minimization *via* accessing the NAB library. In addition, more scoring functions are also introduced in DOCK6 as described in the web manual. Regarding sampling methods, DOCK6 implements several advanced algorithms to study specific interactions between ligands and proteins (e.g., anchor-and-grow, *de novo*, GA, and Hungarian algorithm) [46].

For the purpose of selecting spheres and generating DMS (Database Management System) files to store structure management, DOCK6 needs UCSF Chimera. Docking box construction is dependent on selected spheres, as seen by the UCSF Chimera sphere table. When global docking conditions are met, the protein is covered entirely by the "0" option. For steric overlaps and fast score evaluation, grid creation uses bump and scoring grids, which are modifiable based on the task at hand. Every scoring algorithm operates independently, with results stored in separate files. Ligand energy minimization – which can be adjusted using DOCK6 scripts – is essential to lessen atomic collisions. For best results, flexible ligand docking requires modifying parameters, such as number_scored_conformers and rank_ligands [46]. Long docking periods and resource demands result from the search algorithm's limited ability to manually customize boxes while effectively identifying $3C^{pro}$ binding locations.

The docking tasks in DOCK6 are faster compared to Vina, even when utilizing default settings. The process takes approximately less than 30 minutes for a single system, comprising one protein model and five ligands. Although DOCK6

DOI: http://dx.doi.org/10.5772/intechopen.1005158

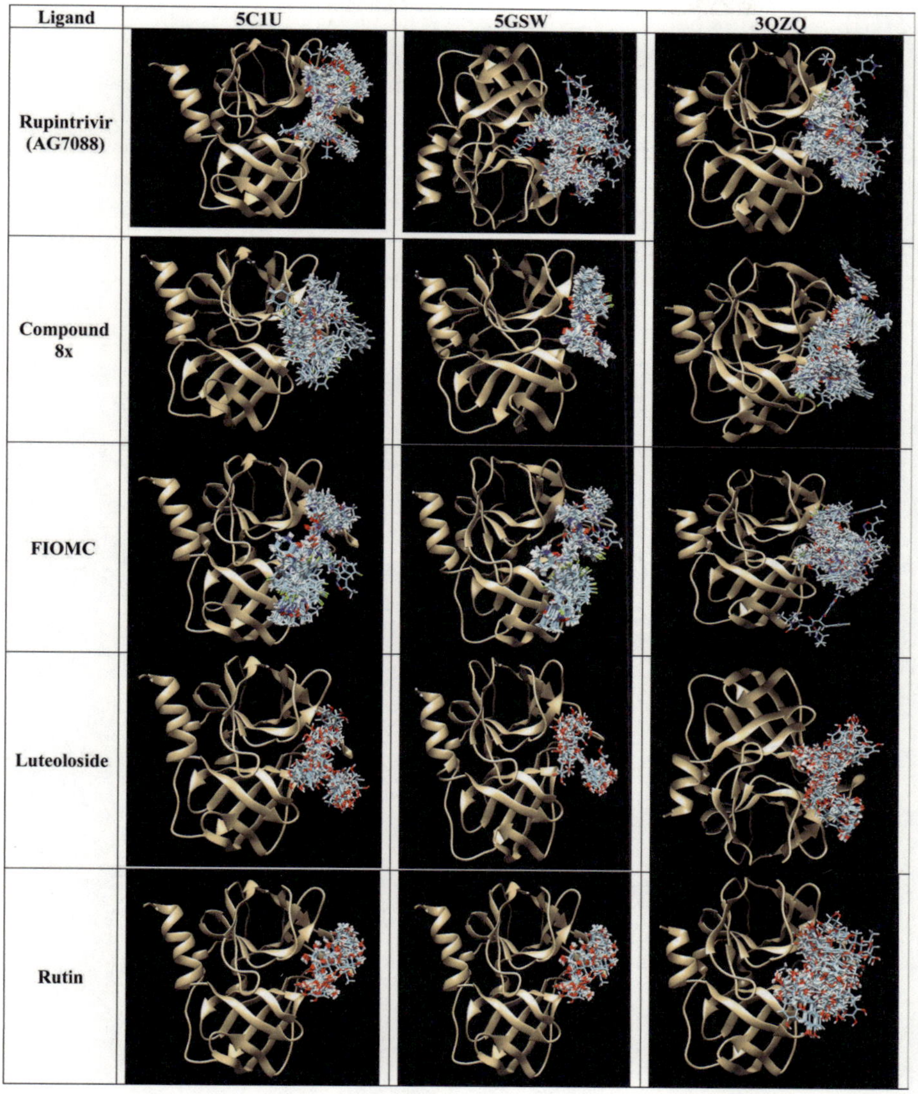

Figure 4.
Binding poses formed by DOCK6 between control ligands (line presentation) on three 3C^{pro} models (ribbon presentation).

is implemented with multiple scoring functions, a default grid score is used in our study case. Furthermore, DOCK6 utilizes an anchor number to determine the generated modes of ligands, making a single run adequate for assessment. Looking at the docking results in **Figure 4**, it is observable that all of ligands interact at the active site of 3C^{pro} protein within three studied models. As a result, DOCK6 is more efficient in defining the active site of the targeted protein *via* evaluating the spheres covered the favorable binding sites in comparison to AutoDock Vina. Furthermore, we notice that rupintrivir binds more accurately at the active site than other ligands, which is similar to the above AutoDock Vina' results. Overall, there is not much different among these models when using DOCK6 for docking.

3.3 MOE

Paul Labute, the President and CEO of Chemical Computing Group, which was established in 1994, is located in Montreal, Quebec, Canada (cite web). The team's goal is to generate innovative tools that not only confront the existing norms but also transform scientific approaches. Thus, the launch of MOE demonstrates the company's dedication to the generation of cutting-edge solutions that remarkably assist in the development of computational chemistry research area. Its operation can carry out in multiple systems (https://www.chemcomp.com/Products.htm).

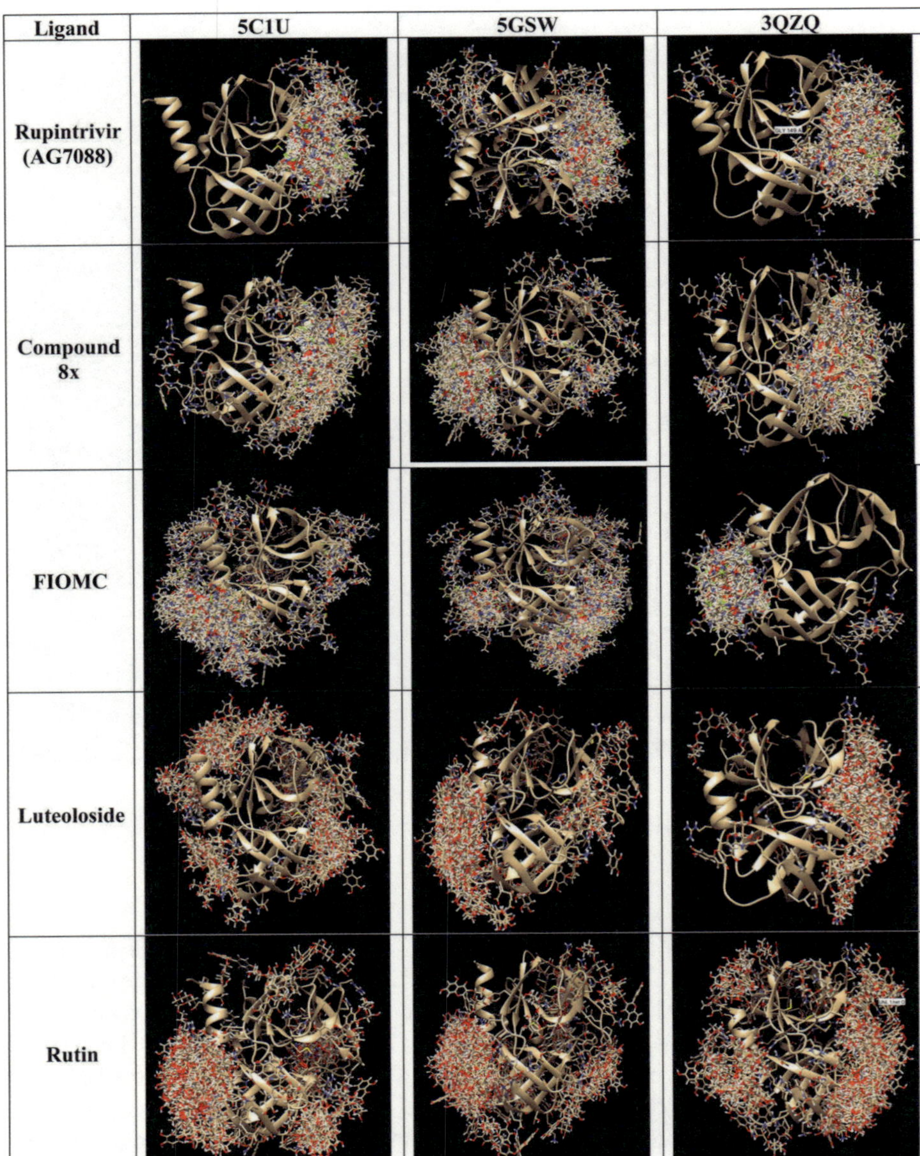

Figure 5.
Binding poses formed by MOE between control ligands (line presentation) on three 3C^{pro} models (ribbon presentation).

In regard to the docking time, it is noticeable that MOE requires the longest time ($\geq$ 1.5 hours) for a pair of protein ligand to produce 100 binding poses. When assessing the percentage of poses at the active site, Rupintrivir ranks the best potent compound in interacting with $3C^{pro}$ proteins at that site, while FIOMC and rutin bind at the opposite site (**Figure 5**).

Moreover, ligands tend to scatter in all cases due to default search algorithm setting. Apart from the wild-type strain, two other mutant strains show opposite results compared to two previous tools. In particular, ligands distributed more neatly in 3QZQ case, compared to 5GSW, where three out of five ligands do not bind at the active site. This is contrast to our earlier study that mutation in 5GSW did not show many impacts on the binding of ligand rather than that of 3QZQ [36].

The rational in choosing the parameter depends on the purpose of docking. Among them, the pharmacophore only supports the commercial version. Meanwhile, the other scoring functions can be changed depending on the interested interactions (e.g., Atom-Sphere Exclusion, Affinity Δ G scoring, Alpha Hydrogen Bond scoring, and GBVI/WSA), alongside with several crucial factors, including atom-atom interactions, geometric fit, H-bonding, and energy contribution. In our study case, default settings are employed for all systems that use triangle matcher in placement; London Δ G in rescoring, and no refinement is set.

4. Conclusion

In conclusion, this study examined several scoring functions and search algorithms in docking software, as well as the field of docking approaches, identifying local and global docking. This discussion centered on global docking, more especially on complete and partial flexible docking. In order to identify protein binding sites, three popular docking tools – AutoDock Vina, MOE, and DOCK6 – were evaluated using a case study containing the EV-A71 3C provirus that causes HFMD. Interestingly, DOCK6 outperformed the others in displaying concentrated ligand poses in the 5C1U protein's active regions, while AutoDock Vina provided information about how mutations affect ligand interactions.

In order to evaluate the quality of docking results and support the logical design of new therapies, post-docking analysis requires accurate binding energy prediction. However, this is not possible at this level of global docking because the main goal is to evaluate receptor binding locations. These binding energy values must be ascertained by additional analyses, such as MD simulation or calculations utilizing the molecular mechanics (MM)/PBSA or MM/GBSA methods. Meanwhile, by utilizing the AMBER score and a basic MM-GB/SA method, DOCK6 assesses binding energy.

Regarding the study case, rupintrivir (AG7088) consistently showed excellent inhibitory activity across both wild-type and mutant $3C^{pro}$ animals, establishing it as a reliable positive control despite the variation in the data. All ligands docked at the active site of $3C^{pro}$ proteins in DOCK6 with the fastest time in less than 5 minutes for each pair of protein ligand when 100 poses are generated. Meanwhile, MOE does not appear to support a comprehensive evaluation of a putative binding site, depending on the quantity of produced ligands posed at various binding sites. Furthermore, the likelihood that ligands will bind to $3C^{pro}$ proteins is affected by mutations in either 5GSW or 3QZQ. As a result, the docking findings produced by MOE show something different from what we saw in AutoDock Vina's previous studies (**Table 3**).

	System requirement	Advantages	Disadvantages
AutoDock Vina	• Support for 64-bit Linux, Mac OS X 10.15 (Catalina) or later versions, SGI IRIX, and Microsoft Windows • Hardware: ≥ 4GB RAM • Disk space: ≥ 100 MB • Compatible with CPUs only • Simple command–line interface: easy use • MGLtools is perquisite for preparing ligand and receptor docking inputs	• Less settings: grid box, exhaustiveness value • Perform docking for multiple protein -ligand systems at once with a Perl script • An open–source software with widely used globally promotes scientific cooperation and continual progress to address any difficulties • Flexible ligand docking options • Academic free	• Ligands and proteins in must PDBQT format, which is not ready to use • Higher exhaustiveness value requires extensive computational resources. • Complicated systems require longer timescale. • Limited side chains of receptors flexibility. • Solvation models by hydrated docking, prefer for AutoDock4 rather than Vina
DOCK6	• Support for Linux or Windows (Unix-like environment – Cygwin); Mac OS with GNU compilers and configuration files • Hardware: ≥ 4GB RAM • Disk space: ≥ 200 MB • CPUs for basic dockings and GPU compiler • Command-line interface: easy to use	• Fast docking process for a default setting • Efficient search algorithms • Decrease sampling • Various scoring functions help reducing scoring failure • Solvation models and flexible receptors docking: tackled by AMBER score • Flexible ligand docking option • Available free	• Limited in the manual customization of the box • Multiple parameter settings to evaluate • Require intensive computational resources for AMBER score docking • Need to use with other programs

	System requirement	Advantages	Disadvantages
MOE	• Support in Windows, Linux, MacOS • Hardware: ≥ 4GB RAM • Disk space: ≥ 1GB • Potential energy computations: 8 CPUs requirement • Command-line and graphical interface unit: easy to use	• Allow multi-tasking with command line • Can customize wall constraint for each protein • Unlimited number of poses per experiment • Support many file formats • Flexibility of receptors with several limited side chains • Flexibility of ligands with ≤ 10 rotatable bonds • Increases adaptability by several search strategies • Maintains original geometry to improve search efficiency and precision	• Uncommon command language SVL • More time is needed for a complex system • More pharmacophore is required for ≥ 10 movable bonds • Solvation models handling for the commercial version solely • Require ring structures to be generated in advance • Commercial tool

Table 3.
An analysis of the three docking systems' respective system requirements, benefits, and drawbacks.

Default settings are frequently used since developers propose them for the purpose of tool comparison. However, as many possibilities have been covered previously, it is crucial to choose settings that are appropriate for the research aims in order to produce the best results. To guarantee a more complete understanding of protein-ligand interactions in the quest for drug discovery and design, researchers may investigate other technologies in the future to improve and build upon current discoveries. Efficient molecular docking for big chemical libraries is made possible by developments in computational resources, such as cloud and distributed computing. Processing operations on GPUs are accelerated, particularly when large conformational landscapes are being explored. Advancements such as cryo-electron microscopy offer abundant three-dimensional information, augmenting understanding of protein structure and maybe boosting docking accuracy. Pose prediction accuracy is increased by methods, including interaction fingerprints, better MM/PBSA analysis, and pharmacophore modeling integration. The development of drug delivery devices is aided in comprehending complex dynamics and binding processes by combining docking with MD simulations. Enhancing docking technologies, such as optimizing metaheuristics for accurate pose prediction, is possible with AI, especially with deep learning and machine learning. By iteratively retraining using informative data, active learning minimizes the need for screening resources while optimizing prediction reliability. Support vector machines and random forests are examples of machine-learning algorithms that outperform traditional score methods in terms of accuracy, which helps in ligand ranking. Drug research and design are improved by deep learning methods, such as convolutional and recurrent neural networks, which extract spatial and sequential information from 3D structures. The combination of AI and ML has the potential to transform docking tools, speed up the hunt for novel therapies, and enable more efficient drug discovery and design than is possible with current methods.

Conflict of interest

The authors declare no conflict of interest.

Other declarations

None.

Author details

Vy T.T. Le[1,2], Tu H.T. Nguyen[1,2] and Phuc-Chau Do[1,2*]

1 School of Biotechnology, International University, Hochiminh City, Vietnam

2 Vietnam National University, Ho Chi Minh City, Vietnam

*Address all correspondence to: dnpchau@hcmiu.edu.vn

References

[1] Stanzione F, Giangreco I, Cole JC. Use of molecular docking computational tools in drug discovery. In: Progress in Medicinal Chemistry. Netherlands: Elsevier B.V; 2021. pp. 273-343

[2] Meng XY, Zhang HX, Mezei M, Cui M. Molecular docking: A powerful approach for structure-based drug discovery. Current Computer-Aided Drug Design. 2011;7(2):146-157

[3] Li D, Ji B. Protein conformational transitions coupling with ligand interactions: Simulations from molecules to medicine. Medicine in Novel Technology and Devices. 2019;**3**:100026

[4] Koshland DE. Correlation of structure an function in enzyme actio theoretical and experimental tools are leading correlations between enzyme structure and function. Science. 1963;**142**:1533-1541

[5] Zhou Y, Jiang Y, Chen SJ. RNA–ligand molecular docking: Advances and challenges. Wiley Interdisciplinary Reviews: Computational Molecular Science. 2022;**12**:e1571

[6] Pagadala NS, Syed K, Tuszynski J. Software for molecular docking: A review. Biophysical Reviews. 2017;**9**:91-102

[7] Kellenberger E, Rodrigo J, Muller P, Rognan D. Comparative evaluation of eight docking tools for docking and virtual screening accuracy. Proteins: Structure, Function and Genetics. 2004;**57**(2):225-242

[8] Wang Z, Sun H, Yao X, Li D, Xu L, Li Y, et al. Comprehensive evaluation of ten docking programs on a diverse set of protein-ligand complexes: The prediction accuracy of sampling power and scoring power. Physical Chemistry Chemical Physics. 2016;**18**(18):12964-12975

[9] Pronk S, Páll S, Schulz R, Larsson P, Bjelkmar P, Apostolov R, et al. GROMACS 4.5: A high-throughput and highly parallel open source molecular simulation toolkit. Bioinformatics. 2013;**29**(7):845-854

[10] Sousa SF, Fernandes PA, Ramos MJ. Protein-ligand docking: Current status and future challenges. Proteins: Structure, Function and Genetics. 2006;**65**:15-26

[11] Dias R, Filgueira De Azevedo W. Molecular docking algorithms. Current Drug Targets. 2008;**9**:1040-1047

[12] Korb O. Efficient ant colony optimization algorithms for structure- and ligand-based drug design. Chemistry Central Journal. 2009;**3**(S1):O10

[13] Abagyan R, Totrov M, Kuznetsov D. ICM-A new method for protein modeling and design: Applications to docking and structure prediction from the distorted native conformation docking and global energy optimization. Journal of Computational Chemistry. 1994;**15**(5):994

[14] Ewing TJ, Makino S, Geoffrey Skillman A, Kuntz ID. DOCK 4.0: Search strategies for automated molecular docking of flexible molecule databases. Journal of Computer-Aided Molecular Design. 2001;**15**:411-428

[15] Trosset JY, Scheraga HA. PRODOCK: Software package for protein modeling and docking. Journal of Computational Chemistry. 1999;**20**(4):412427

[16] Pierce BG, Wiehe K, Hwang H, Kim BH, Vreven T, Weng Z. ZDOCK server: Interactive docking prediction of protein-protein complexes and symmetric multimers. Bioinformatics. 2014;**30**(12):1771-1773

[17] Macindoe G, Mavridis L, Venkatraman V, Devignes MD, Ritchie DW. HexServer: An FFT-based protein docking server powered by graphics processors. Nucleic Acids Research. 2010;**38**(SUPPL. 2):W445-W449

[18] Venkatraman V, Yang YD, Sael L, Kihara D. Protein-protein docking using region-based 3D Zernike descriptors. BMC Bioinformatics. 2009;**10**:407

[19] Huang SY, Zou X. Advances and challenges in protein-ligand docking. International Journal of Molecular Sciences. 2010;**11**:3016-3034

[20] Allen WJ, Balius TE, Mukherjee S, Brozell SR, Moustakas DT, Lang PT, et al. DOCK 6: Impact of new features and current docking performance. Journal of Computational Chemistry. 2015;**36**(15):1132-1156

[21] Verdonk ML, Ludlow RF, Giangreco I, Rathi PC. Protein-ligand informatics force field (PLIff): Toward a fully knowledge driven "force field" for biomolecular interactions. Journal of Medicinal Chemistry. 2016;**59**(14):6891-6902

[22] Jain AN. Surflex: Fully automatic flexible molecular docking using a molecular similarity-based search engine. Journal of Medicinal Chemistry [Internet]. 2003;**46**(4):499-511. DOI: 10.1021/jm020406h

[23] Huang SY, Grinter SZ, Zou X. Scoring functions and their evaluation methods for protein-ligand docking: Recent advances and future directions. Physical Chemistry Chemical Physics. 2010;**12**(40):12899-12908

[24] Kitchen DB, Decornez H, Furr JR, Bajorath J. Docking and scoring in virtual screening for drug discovery: Methods and applications. Nature Reviews Drug Discovery. 2004;**3**:935-949

[25] Li J, Fu A, Zhang L. An overview of scoring functions used for protein–ligand interactions in molecular docking. Interdisciplinary Sciences – Computational Life Sciences. 2019;**11**:320-328

[26] Eberhardt J, Santos-Martins D, Tillack AF, Forli S. AutoDock Vina 1.2.0: New docking methods, expanded force field, and python bindings. Journal of Chemical Information and Modeling. 2021;**61**(8):3891-3898

[27] Trott O, Olson AJ. AutoDock Vina: Improving the speed and accuracy of docking with a new scoring function, efficient optimization, and multithreading. Journal of Computational Chemistry. 2009;**31**:455-461

[28] Molecular Operating Environment (MOE). [Internet]. 2022.02 Chemical Computing Group ULC 910-1010 Sherbrooke St. W., Montreal, QC, Canada. 2024. Available from: https://www.chemcomp.com/ [Accessed: February 27, 2024]

[29] Koh WM, Badaruddin H, La H, Chen MIC, Cook AR. Severity and burden of hand, foot and mouth disease in Asia: A modelling study. BMJ Global Health. 2018;**3**(1):e000442

[30] Ong KC, Wong KT. Understanding enterovirus 71 neuropathogenesis and its impact on other neurotropic enteroviruses. Brain Pathology. 2015;**25**(5):614-624

[31] Nayak G, Bhuyan SK Bhuyan R, Sahu A, Kar D, Kuanar A. Global emergence of enterovirus 71: A systematic review. Beni-Suef University Journal of Basic and Applied Sciences. 2022;**11**:78, 11 pages

[32] Hu K, Onintsoa Diarimalala R, Yao C, Li H, Wei Y. EV-A71 mechanism of entry: Receptors/Co-receptors, related pathways and inhibitors. Viruses. 2023;**15**:785

[33] Cui S, Wang J, Fan T, Qin B, Guo L, Lei X, et al. Crystal structure of human enterovirus 71 3C protease. Journal of Molecular Biology. 2011;**408**(3):449-461

[34] Sun D, Chen S, Cheng A, Wang M. Roles of the picornaviral 3c proteinase in the viral life cycle and host cells. Viruses. 2016;**8**(3):82

[35] Wang J, Fan T, Yao X, Wu Z, Guo L, Lei X, et al. Crystal structures of enterovirus 71 3C protease complexed with rupintrivir reveal the roles of catalytically important residues. Journal of Virology. 2011;**85**(19):10021-10030

[36] Le TTV, Do PC. Molecular docking study of various enterovirus—A71 3C protease proteins and their potential inhibitors. Frontiers in Microbiology. 2022;**13**:987801

[37] Lu G, Qi J, Chen Z, Xu X, Gao F, Lin D, et al. Enterovirus 71 and coxsackievirus A16 3C proteases: Binding to rupintrivir and their substrates and anti-hand, foot, and mouth disease virus drug design. Journal of Virology. 2011;**85**(19):10319-10331

[38] Morris GM, Ruth H, Lindstrom W, Sanner MF, Belew RK, Goodsell DS, et al. Software news and updates AutoDock4 and AutoDockTools4: Automated docking with selective receptor flexibility. Journal of Computational Chemistry. 2009;**30**(16):2785-2791

[39] Santos-Martins D, Solis-Vasquez L, Tillack AF, Sanner MF, Koch A, Forli S. Accelerating AutoDock 4 with GPUs and gradient-based local search. Journal of Chemical Theory and Computation. 2021;**17**(2):1060-1073

[40] Ravindranath PA, Forli S, Goodsell DS, Olson AJ, Sanner MF. AutoDockFR: Advances in protein-ligand docking with explicitly specified binding site flexibility. PLoS Computational Biology. 2015;**11**(12):e1004586

[41] DesJarlais RL, Sheridan RP, Dixon JS, Kuntz ID, Venkataraghavan R. Docking flexible ligands to macromolecular receptors by molecular shape. Journal of Medicinal Chemistry. 1986;**29**(11):2149-2153

[42] Prentis LE, Singleton CD, Bickel JD, Allen WJ, Rizzo RC. A molecular evolution algorithm for ligand design in DOCK. Journal of Computational Chemistry. 2022;**43**(29):1942-1963

[43] Brozell SR, Mukherjee S, Balius TE, Roe DR, Case DA, Rizzo RC. Evaluation of DOCK 6 as a pose generation and database enrichment tool. Journal of Computer-Aided Molecular Design. 2012;**26**:749-773

[44] Allen WJ, Fochtman BC, Balius TE, Rizzo RC. Customizable de novo design strategies for DOCK: Application to HIVgp41 and other therapeutic targets. Journal of Computational Chemistry. 2017;**38**(30):2641-2663

[45] Lang PT, Brozell SR, Mukherjee S, Pettersen EF, Meng EC, Thomas V, et al. DOCK 6: Combining techniques to model RNA-small molecule complexes. RNA. 2009;**15**(6):1219-1230

[46] William Joseph Allen (TACC), Trent Balius (FNLCR), John Bickel (SUNY-Stony Brook), Brock Boysan (SUNY-Stony Brook), Scott Brozell (Rutgers University), Chris Corbo (SUNY-Stony Brook), et al. DOCK 6.11 Users Manual [Internet]. 2023. Available from: https://dock.compbio.ucsf.edu/DOCK_6/dock6_manual.htm [Accessed: February 27, 2024]

Chapter 4

Interplay of Epitope Landscapes: Unraveling Molecular Recognition through Mapping, Docking, and Simulation

Shine P. Varkey and Shankar K.M.

Abstract

Molecular docking is a crucial process in computational immunology that involves the precise binding of antibodies to antigens, illuminating the key sites responsible for immune recognition named epitopes. Understanding these molecular interactions at a detailed level is critical in unraveling immune responses and developing targeted interventions. Commencing with the intricacies of antibody design, the review will explore the connection among molecular features, including interaction environments, structural features, and the specificity of these interactions. The review will underscore how this approach not only expedites research processes but also offers cost-effective avenues for in-depth exploration of molecular landscapes. Finally, the practical applications of these insights in diagnostics and vaccine development will be discussed, showcasing the transformative potential of understanding and manipulating antibody-antigen interactions at the molecular level.

Keywords: antigen-antibody interactions, epitope landscape, antibody modeling, computational docking, interaction environment, diagnostic and therapeutic development

1. Introduction

The field of antibody production and engineering has advanced significantly, enabling the development of monoclonal antibodies tailored for specific therapeutic targets. High-throughput sequencing technologies have revolutionized our understanding of the B-cell receptor repertoire, providing insights into antibody diversity and facilitating the design of antibodies with enhanced binding affinity and specificity [1]. This progress has been instrumental in the production of antibodies for a variety of applications, from cancer immunotherapy to the treatment of autoimmune diseases. At the heart of antibody-antigen interactions lies the concept of epitope mapping, the process of identifying the precise binding sites on an antigen where an antibody attaches. This understanding is crucial for the design of vaccines and therapeutic antibodies as it allows for the precise targeting of pathogens or diseased cells. Epitope mapping not only aids in the development of targeted interventions but also enhances our understanding of immune recognition and response mechanisms.

Despite the advancements in sequencing and antibody engineering, accurately predicting the binding sites of antibodies on antigens remains a complex challenge. Traditional experimental methods, such as X-ray crystallography and mutagenesis, have provided valuable insights but are often time-consuming, expensive, and limited in their throughput. Computational methods offer a promising alternative, capable of high-throughput analyses and providing atomic-level details of antibody-antigen interactions [1, 2].

Molecular docking emerges as a powerful computational tool in this context, predicting how an antibody interacts with its antigen [3]. The development of docking algorithms, such as those implemented in ClusPro, has significantly advanced our ability to predict these interactions accurately. These tools use sophisticated models to simulate the physical forces between the antibody and antigen, providing valuable insights into the nature of their binding. The introduction of advanced computational platforms, such as ClusPro-AbEMap, represents a significant leap forward in epitope mapping [4]. By integrating template-based modeling with docking algorithms, these tools can predict epitope sites with unprecedented accuracy. This approach not only accelerates the discovery of potential therapeutic targets but also opens new avenues for vaccine development and the study of immune responses.

Molecular docking stands as a cornerstone in the interdisciplinary field of computational biology and immunology, offering a simulation-based approach that predicts the optimal interaction between two molecules—typically, a smaller molecule (ligand) and a larger molecule (receptor), such as an antibody or enzyme [3]. In computational immunology, this technique is pivotal for simulating the complex dance between antibodies (the defenders of the biological realm) and antigens (foreign invaders such as viruses and bacteria). This interaction is fundamental for identifying how the immune system recognizes, attaches to, and neutralizes pathogens. By simulating these interactions, molecular docking allows scientists to predict the binding affinity between molecules, providing insights into the structural configurations that underlie immune recognition and response mechanisms.

The origin of molecular docking can be traced back to the early 1980s, a period marked by increasing interest in leveraging computational power to solve biological problems [5]. Initially, the practice faced significant hurdles, including limited computational resources and a nascent understanding of molecular dynamics and interactions. However, as computational technology advanced, so too did molecular docking techniques, evolving from simplistic models to sophisticated algorithms capable of simulating the dynamic nature of molecular interactions with remarkable accuracy. This evolution was propelled by the advent of high-throughput computing and enhanced algorithms for predicting molecular behavior, marking a paradigm shift in how researchers approached the study of molecular interactions in variety of areas [6].

The principal aim of molecular docking in immunology is to elucidate the interactions between antibodies and antigens at a granular level [7]. This insight is critical for several reasons. First, it sheds light on the immune system's recognition and neutralization pathways, revealing how antibodies can be engineered or elicited through vaccination to target specific pathogens effectively. By understanding these molecular dialogs, scientists can identify the most accessible and immunogenic parts of an antigen, guiding vaccine design and antibody therapy development.

Furthermore, molecular docking has revolutionized the field of drug discovery [6]. Its ability to screen vast compound libraries for potential interactions with biological targets offers a faster, more cost-effective route to identifying new drugs. This is particularly crucial in the rapid response to emerging infectious diseases, where time is of the essence. The predictive power of molecular docking has facilitated the development of targeted therapies, contributing significantly to personalized medicine by tailoring treatments to the individual's specific molecular profile.

2. Understanding epitopes: types, features, and mapping

Epitopes represent the cornerstone of immune recognition, driving the development of targeted immunotherapies and diagnostic strategies. Understanding the diverse types and features of epitopes, coupled with sophisticated mapping techniques, empowers researchers in the quest for effective vaccines and immunotherapies against infectious diseases, autoimmune disorders, and cancer. Continued advancements in epitope research hold promise for personalized medicine and precision immunotherapy in the future.

2.1 Types of epitopes

2.1.1 Continuous epitopes

A linear peptide fragment that binds to the antibody is called a continuous epitope [1]. Since the classification criterion is based on binding activity, the continuous epitope is defined functionally, and there are no reports suggesting the role of every residue in the peptide making contact with the antibody. The binding of individual amino acids in the peptide epitope is determined by positional scanning with amino acid residue replacement [8]. Despite their rare occurrences, linear epitopes find potential use as reagents alternative to infectious agents in diagnostics and as immunogens in developing antibody reagents and synthetic vaccines overall [9].

2.1.2 Discontinuous epitopes

The second category of epitopes, called discontinuous or conformational epitopes, are formed by the amino acid residues brought together by protein folding [1]. The majority of the epitopes studied are discontinuous in nature, and their interaction with the antibody requires the folded form of the native protein. For this reason, discontinuous epitopes cannot be extracted as a separate segment from the original protein molecule nor is it possible to demonstrate their binding activity independently. Apart from practical use, this phenomenon also limits the functional assignment of the epitope as it can only be defined structurally, and mapping studies require sophistication, such as X-ray crystallography or NMR spectroscopy.

2.2 Features of epitopes

Epitopes play crucial roles in immune recognition and response. Firstly, they exhibit antigenic specificity, selectively binding to complementary antibodies or T-cell receptors, which forms the foundation for immune recognition and the generation of targeted immune responses against pathogens. Additionally, epitopes possess immunogenicity, the ability to induce an immune response. Several factors influence immunogenicity, including epitope accessibility, structural stability, and the host immune repertoire [7]. Moreover, epitopes demonstrate flexibility and plasticity, enabling them to accommodate structural variations without compromising antibody binding. This flexibility facilitates antigenic variation in pathogens, aiding immune evasion strategies. Finally, epitopes may display cross-reactivity, where antibodies raised against one epitope recognize structurally similar epitopes on unrelated antigens. This cross-reactivity has implications for vaccine design and diagnostic assays, highlighting the importance of careful epitope selection to minimize off-target effects.

2.3 Mapping techniques for epitopes

2.3.1 Experimental approaches

X-ray crystallography determines the three-dimensional structure of antigen-antibody complexes at atomic resolution, offering insights into conformational epitopes [10]. Nuclear magnetic resonance (NMR) spectroscopy studies protein structure and dynamics in solution, useful for mapping flexible epitopes [11]. Surface plasmon resonance (SPR) quantitatively analyzes epitope-paratope binding kinetics and affinity, providing real-time insights into molecular interactions [12].

2.3.2 Computational approaches

Homology modeling predicts epitope structures based on sequence similarity, aiding in peptide vaccine design. Molecular docking epitope binding modes and affinities, facilitating virtual screening of epitope libraries.

3. Antibody-antigen interaction and interfacial properties

3.1 Amino acid composition and propensity

In the complex interplay between antibodies and antigens, the amino acid composition of their interface plays a critical role in determining binding specificity and affinity. Analyzing the amino acid composition of both the epitope contact surface (ECS) of antigens and the paratope contact surface (PCS) of antibodies unveils key insights into their structural and functional characteristics [13]. The features of antibody-binding interfaces, such as size, shape, and chemical composition, have been analyzed extensively [13, 14]. Studies suggest that the interfaces of antigen-antibody complexes differ significantly from those of nonimmune protein complexes. These interfaces are particularly rich in aromatic residues, such as Tyr, Arg, His, Phe, and Trp. The role of enriched aromatic residues, especially Tyr and Trp, in antibody-antigen binding selectivity and specificity has garnered considerable interest. Mutagenesis studies have highlighted the unpredictable nature of affinity changes upon mutating these residues, with some identified as "hot spots" contributing significantly to high-affinity interactions [13]. Tyr and Trp are noted for their ability to form close interactions with antigen proteins due to their aromatic and hydroxyl groups. This contrasts with Phe, another aromatic residue, which is rarely found at antibody interfaces, possibly due to the absence of functional groups in its aromatic ring. The piling up of aromatic rings in these "aromatic islands (AI)" may provide extra stabilizing energy to protein complexes, creating a highly hydrophobic environment that favors hydrophobic interactions [14]. This dense gathering of aromatic side chains contributes significantly to antigen-antibody binding. It is also suggested that cooperative effects exist within AI, leading to an increase in binding affinity with more Tyr or Trp residues stacking together. Mutations of aromatic residues in aromatic islands can have significant effects on binding affinity, as suggested by studies [15].

The ECS of antigens exhibits a diverse range of amino acid residues, with peptide antigens displaying a smaller and more variable composition compared to larger protein antigens [13]. Interestingly, the PCS of antibodies contains almost twice the number of amino acid residues as the ECS, indicating a higher density of interaction. Furthermore, certain residues, such as TrpP and TyrR, are enriched in the interface, suggesting their pivotal role in the recognition process. The occurrence propensity of amino acids in the interface normalizes their tendencies to occur on protein surfaces and provides a measure of their significance in antibody-antigen interactions. Residues, such as TrpP and TyrR, exhibit higher occurrence propensity in the ECS, indicating their importance in driving specific interactions. Conversely, residues, such as CYS and Phe, show lower occurrence propensity, suggesting their lesser involvement in interface interactions.

3.2 Interaction specificity and substitutability

The specificity of interactions between amino acid residues in the antibody-antigen interface is crucial for understanding the molecular basis of binding affinity and selectivity. Analysis of interaction frequency and substitutability gives insights into the complex nature of these interactions and their functional implications.

Certain amino acid residues exhibit high substitutability, suggesting their roles in maintaining interface integrity and specificity [13]. By assessing the correlation between pairs of residues, it becomes evident which residues are more likely to interact with specific counterparts on the opposite surface. Residues with low correlation values indicate lower substitutability, while those with higher correlation values suggest higher substitutability and possibly more promiscuous interactions. Although the number of contacts in specific regions of antibodies can be mapped, it does not directly correlate with the energetic contributions to the binding interface [16]. However, regions such as CDRH3, known for mediating antibody contacts, show high populations of unique paratope residues (UPRs) suggesting a potential correlation between the number of UPRs and energetic contributions. Noncontact residues in antibodies may play a significant role in supporting the orientation and flexibility of paratope residues.

These findings provide valuable insights into the structural determinants of antibody-antigen recognition and have implications for engineering antibodies with enhanced binding properties. Understanding the specificity and substitutability of interface residues enables the design of antibodies with tailored binding affinities for therapeutic and diagnostic applications.

3.3 Spatial distribution and functional implications

The spatial distribution of amino acid residues within the antibody-antigen interface offers valuable insights into the structural organization and functional implications of different regions [13, 17]. Analysis of residue distribution unveils patterns that reflect the dynamic nature of interface interactions and their functional consequences. Residues on the PCS are distributed more evenly across the interface compared to those on the ECS, suggesting a dynamic role in mediating interactions with diverse epitopes. Certain residues, such as TyrR, exhibit preferential localization closer to the center of the interface, indicating their involvement in key interactions critical for stabilizing the complex.

Understanding the spatial distribution of residues is essential for deciphering the structural dynamics and functional implications of the antibody-antigen interface. By elucidating these aspects, researchers can gain deeper insights into the mechanisms driving antibody-antigen interactions and develop novel strategies for modulating immune responses and designing therapeutic interventions.

3.4 Salt bridges

Electrostatic interactions have major influence in protein interactions, significantly contributing to binding free energy [18]. Among these interactions, salt bridges, formed by oppositely charged residues pairing up (e.g., positively charged lysine with negatively charged aspartate), play a crucial role. Numerous investigations have underscored the pivotal role of salt bridges in binding affinity. For instance, in the NC10 Fab-NA complex, the presence of an AspH56-LysN432 salt bridge was identified [19]. Disruption of this salt bridge, along with two hydrogen bonds from SerN370 and SerN372, through the replacement of AspH56 with serine, alanine, or glycine, led to a significant loss in binding, underscoring the essential role of salt bridge formation. In a separate study aimed at enhancing the binding affinity of AspH56, replacement with glutamate was explored [20]. It was found that the terminal side chain carboxyl oxygen of GluH56 was brought into closer proximity to LysN432, thereby strengthening the salt bridge. However, this replacement resulted in a substantial 40-fold decrease in binding affinity, presumably due to steric clashes introduced by the bulkier side chain, hindering the formation of binding energies. Studies estimating the contribution of ΔG 5–7.0 kcal/mol energy by the charged Asp H52 further underscore its significance [21], suggesting its essential role in the functional epitope region, which encompasses several charged residues [22]. This functional significance of salt bridges is reiterated in various other studies [23, 24].

3.5 Hydrogen bonds in antibody binding

Despite assumptions regarding the heightened significance of hydrogen bonding in achieving specificity and affinity, discrepancies arise regarding their precise energy contribution values [25–27] primarily due to the presence of water molecules competing for hydrogen bonding sites. Interfacial water molecules emerge as key contributors to hydrogen bond formation in numerous protein-protein complexes [28]. In a recent investigation, hydrogen bonds within the HyHEL-10-HEL complex were found to be either buried or partially exposed [29]. Moreover, deletion of hydrogen bonds resulted in thermodynamic compensation for structural rearrangements caused by mutations.

Consistent with observations in various biomolecular complexes, unfavorable enthalpy changes are compensated by favorable entropy changes [30, 31], suggesting a degree of tolerance for the loss of hydrogen bonds within antigen-antibody complexes. Interfacial water molecules play a significant role in stabilizing antigen-antibody interactions by facilitating hydrogen bond formation. Additionally, hydrogen bonds are reported to reduce conformational flexibility, leading to the stiffening of antigen-antibody complexes [32, 33]. While interfacial water molecules are proposed to enhance the specificity of protein-protein interactions [34], their energy contributions have been studied in only a limited number of protein complexes [35, 36].

3.6 Shape complementarity

Molecular recognition relies heavily on the complementarity of shapes at protein interfaces (**Figure 1**). Previous studies have analyzed shape complementarity in protein interfaces, finding that antibody-antigen interfaces exhibit lower average shape complementarity compared to general protein-protein interfaces [37]. This difference is attributed to the independent evolution of antibodies and their antigens, whereas other protein interactions typically evolve together to optimize complementarity. However, later analyses suggested that the lower shape complementarity observed in antibody-antigen complexes may be due to limitations in the quality of crystal structures used [38].

As computational design aims to create novel interfaces by optimizing a single protein, similar to antibodies, understanding the level of shape complementarity achievable in such single-sided interface design is crucial [39]. Previous studies on protein-protein interfaces have shown that obligate interfaces, which are more hydrophobic, exhibit better shape complementarity, and evolve slower than transient interfaces. Protein-protein interfaces are also characterized by an enrichment of non-polar and aromatic residues, with a depletion of charged residues, although arginine residues are a common exception and are often found in interface "hot spots" along with tryptophan and tyrosine residues. Further, analyses have identified residue-level local network patterns in protein-protein interfaces and have identified numerous atomic-resolution motifs in these interfaces [39].

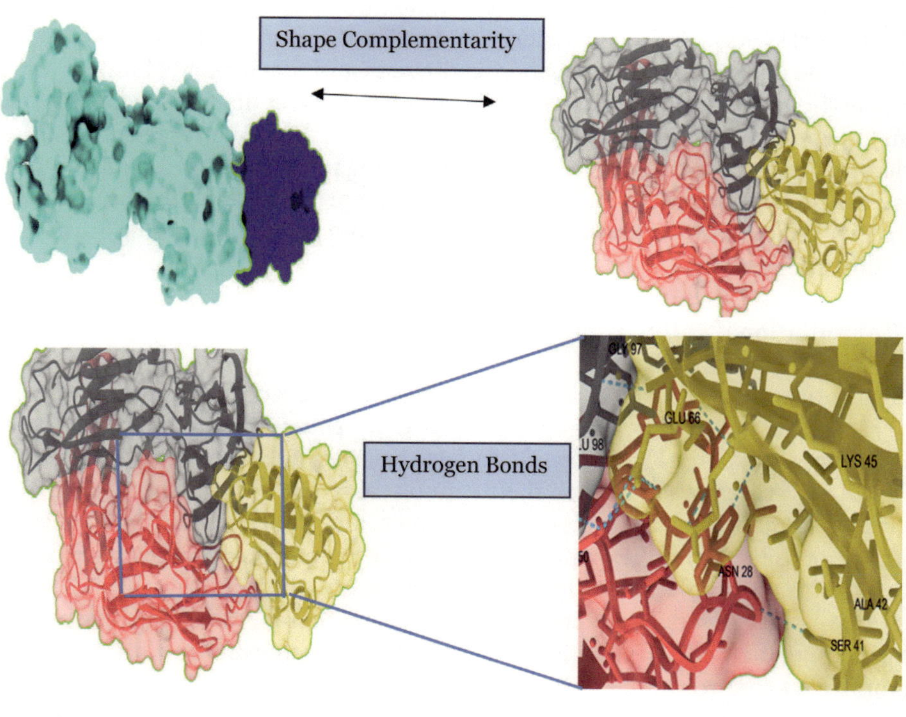

Figure 1.
Shape complementarity and hydrogen bonds influencing antigen-antibody complex formation at the interfacial level.

4. Computational study of antigen-antibody interactions

The computational exploration of antigen-antibody interactions is a fundamental pursuit in immunological research, shedding light on the intricate dynamics associated. Research by Tsumoto and Caaveiro [40] underscores the fundamental role of antigenic nature in shaping the antigen combining site of antibodies. This phenomenon manifests distinctly across antibody classes: haptens elicit deep pocket formations, peptides evoke grooves, while protein binding elicits flat combining sites.

The interplay between the length and residue composition of complementarity determining regions (CDRs) and antigenic properties has garnered considerable attention. Al-Lazikani et al. [41] and Morea et al. [42] elucidated how the six hypervariable loops adopt characteristic conformations dictated by amino acid residues and length, termed canonical conformations. Furthermore, the compositional bias of CDR residues shapes the architecture of the antibody combining site. Residues such as methionine, isoleucine, and serine abound, while phenylalanine, cysteine, valine, and glycine are sparse. Notably, aromatic residues, such as tryptophan and tyrosine, confer a sticky nature to the combining site, crucial for molecular interactions [43, 44]. Distinct antigen-binding classes exhibit varying residue preferences, indicative of their binding specificities. For instance, alanine is underrepresented among hapten binders, while phenylalanine shows scarcity among peptide and hapten binders [45]. Interestingly, certain trends, such as the prevalence of arginine among nucleotide binders, are attributed to interactions with negatively charged phosphate backbones [43].

The significance of residues, such as tyrosine, lies in their multiple contact sites and broad interactive potential with epitopes, facilitating epitope-paratope interactions [13]. Moreover, hydrophobic aromatic residues, such as tyrosine and tryptophan, likely engage in interactive charged group interactions, enhancing specificity and affinity [13, 43]. Hydrogen bonds, van der Waals forces, salt bridges, and hydrophobic interactions collectively contribute to antigen-antibody binding affinity. While hydrogen bonds and van der Waals forces primarily enhance specificity, shape complementarity between binding surfaces and the exclusion of solvent molecules underscore the role of hydrophobic forces in driving interactions [40]. Understanding these molecular nuances not only advances our comprehension of immune responses but also paves the way for innovative therapeutic interventions targeting antigen-antibody interactions.

4.1 Computational antibody modeling

The antigenic interaction of antibodies hinges upon six critical CDRs: L1, L2, L3, H1, H2, and H3, with the hypervariable loops' architecture shedding light on their interactions. Except for H3, all five CDRs have been assigned conformations known as "canonical structures" based on their amino acid sequences and lengths [41, 46]. Evolved through VDJ gene segment arrangements and somatic mutations, H3 loop structures exhibit indefinite conformations, adding complexity to structural determinations, with no established H3 loop structures to date [46, 47].

H3 loops vary in sequence, length, and structure, thereby acquiring diverse antigenic specificities. These regions are pivotal in antigenic recognition, adopting numerous conformations upon antigenic binding [47]. With the emergence of high-resolution antibody crystal structures, a new scheme proposed by [48] aimed to cluster CDR conformations based on conformational energies. This approach eliminated low-resolution CDR loops with high B-factors or high-energy backbone conformations suggested [46]. Utilizing sequence-to-structure relationships derived empirically, it is possible to model L1, L2, L3, H1, and H2 structures with high accuracy, with the root mean square deviation (RMSD) of the modeled structure's backbone to the crystal structure falling below 1.0 Å.

Situated at the core of other loop structures, H3 loops play a significant role in antigen recognition, undergoing substantial mutations during the affinity maturation process compared to other loop structures [49]. They assume an infinite number of conformations, making structure prediction a daunting task, especially during antigenic binding, where they undergo significant structural rearrangements. Computational and experimental studies have built several models based on sequence-to-structure relationships, considering the behavior of H3 loops [50–52].

While H3 loops exhibit indefinite conformations, their C-terminal base regions form limited structures, often in "kinked" or extended forms. These sequence-to-structure rules simplify the challenge of H3 loop modeling by reducing the computational search space [53]. These strategies typically search for appropriate loop conformations from a set of existing loop conformations based on energy or scoring function methods derived from a crystal structural database search [54]. Another approach, adopted by the Rosetta antibody modeling server, combines fragment assembly with a cyclic coordinate descent approach and minimizes fragments using the Rosetta protocol [55]. While the antibody models generated are optimal for further docking analyses in the RosettaDock program [56, 57], the modeling protocols reportedly have limitations in H3 modeling.

In summary, the prediction becomes challenging with longer loop structures, exacerbated by larger flexibility in long surface loops and consequent conformational complexities during antigenic interactions or artifacts arising from crystal packing [58]. Several web servers are available for antibody modeling (**Table 1**) with varying features. Rosetta antibody modeling server, developed by Sircar et al. [59], employs an energy function and rigid body minimization-based VL-VH docking methodology to improve the orientations of H3 and other loops. The Web Antibody Modeling (WAM) server, as proposed by Whitelegg and Rees [60], follows a different approach.

The orientation of variable domains exhibits variability across different antibodies and antigenic interactions [47], impacting the surface involved in antigenic interaction when the structure is inferred from the sequence [39]. Studies examining residue properties and positions across various antibody sequences [61, 62] have revealed that variable domain orientation is sequentially encoded, although this reliance has proven elusive. Furthermore, the domain orientation can vary upon antigenic binding, emphasizing the importance of considering predicted domain orientation when selecting templates for antibody modeling [63]. Covariation analyses of residues representing the Ig-fold, derived from sequence repositories, have unveiled the presence of residue conservation networks in VH and VL domains [64]. These conserved residue network clusters were notably observed at the VL-VH and VH-CH domain interfaces, whereas the VL-CL domain interface lacked such conserved residue networks.

4.2 Computational docking of antibodies

Computational docking antibodies offer a timely and cost-effective alternative to experimentally deduced structures, which often require significant time, expertise, and expense for the crystal formation of complexes. This methodology involves modeling antibody structures and employing computational docking protocols to

Tool/ platform	Methodologies	Features	Advantages and disadvantages
Rosetta antibody	Multistep protocol for modeling the variable domain, template selection, loop grafting, and de novo loop modeling.	Comprehensive modeling protocol for antibody variable domains Integration of template-based modeling and de novo techniques.	Advantages: offers a comprehensive approach for modeling antibody structures Enables accurate prediction of variable domain conformations. Disadvantages: limited availability of templates, especially for nonhuman antibodies may require manual template selection for improved accuracy
AbPredict	Utilizes low-energy combinations of backbone fragments from experimentally determined structures.	Generates antibody models based on low-energy backbone fragment combinations.	Advantages: demonstrates promising performance in modeling antibody structures Disadvantages: utilizes experimentally determined backbone fragments from related structures for modeling
RosettaCM	Multi-template homology modeling protocol suitable for antibodies with noncanonical features or unusual loop lengths	Offers an alternative approach for antibody structure prediction Advantageous for modeling antibodies with nonstandard features.	Advantages: versatile modeling protocol for diverse antibody structures, effective for modeling antibodies with noncanonical features Disadvantages: limited by the availability and quality of homology templates may require optimization for specific antibody structures
WAM Server	Web-based antibody modeling platform utilizes comparative modeling techniques. Integrates experimental data and machine learning algorithms.	Provides a user-friendly interface for antibody modeling incorporates experimental data to enhance accuracy offers rapid and automated modeling of antibody structures.	Advantages: accessibility through a web interface-integration of experimental data improves modeling accuracy. Disadvantages: limited customization compared to standalone software. Dependency on server availability and processing speed.

Table 1.
Overview of the methodologies, features, and advantages and disadvantages of antibody modeling tools and platforms.

establish interaction patterns [65]. While computational docking is continuously improving, limitations arise when using modeled antigen and antibodies as starting structures [66]. However, combining computational predictions with experimental knowledge from epitope mapping or mutagenesis can enhance accuracy in understanding antigen-antibody interactions.

The Rosetta modeling program, as described by Sivasubramanian et al. [65], utilizes existing crystal structures as templates to assign framework regions and construct the beta barrel. It employs loop grafting and loop modeling to achieve non-H3 loop conformations, followed by refinement through Monte Carlo minimization and "all-atom energy function" algorithms. Model building error checking involves two tests: native loop recovery test and homology modeling test. These tests ensure fidelity to crystallographic CDR loop conformation and the ability to build homology models from sequence alone, respectively.

The effectiveness of protein-protein docking in studying antibody-antigen complexes was demonstrated using a modeled antibody and antigen crystal structure

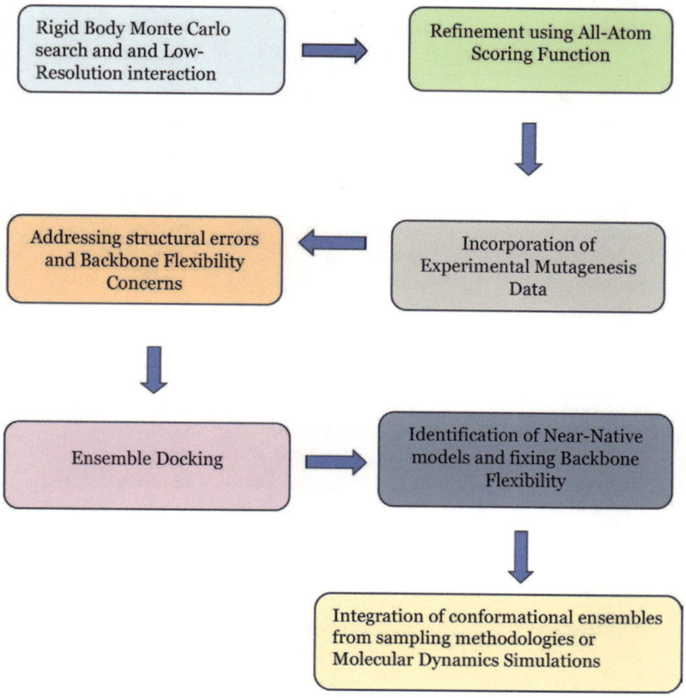

Figure 2.
The RosettaDock scheme illustrates the computational docking process for antibodies, showcasing the workflows of the Rosetta modeling program and ensemble docking.

with the RosettaDock program [55, 65]. The protocol (**Figure 2**) involves rigid body Monte Carlo search and low-resolution interaction potential, followed by refinement using an all-atom scoring function. This approach was applied to study the epidermal growth factor receptor (EGFR) antibody-antigen complex, incorporating experimental mutagenesis data. Structural errors in homology-based models can lead to inaccuracies in subsequent protein-protein docking protocols, especially concerning backbone flexibility during interactions [67]. Ensemble docking, which incorporates conformational ensembles from sampling methodologies or molecular dynamics simulations, addresses this concern efficiently. Ensemble docking has demonstrated success in identifying near-native models and fixing backbone flexibility [68, 69].

Among the various docking algorithms (**Table 2**), RosettaDock has emerged as a prominent tool for modeling antibody-antigen interactions. RosettaDock [70] employs a sophisticated protocol involving low-resolution rigid body perturbations and high-resolution sidechain and backbone minimization to generate models of the docked complex. The algorithm's success lies in its ability to explore a vast conformational space and identify low-energy structures corresponding to native conformations. Benchmarking studies have validated the efficacy of RosettaDock in recovering native conformations, with encouraging results in both local and global searches. RosettaDock is a widely used docking algorithm that enables the prediction of antibody-antigen complexes with full backbone flexibility. The protocol involves low-resolution docking steps followed by high-resolution refinement, allowing for the exploration of conformational space and optimization of side-chain interactions. Despite its effectiveness, RosettaDock relies on the input structures of both docking

Algorithm name	Key features	Accuracy	Computational requirements	Applications
ClusPro	Utilizes sophisticated models for accurate prediction of interactions. Incorporates energy minimization and clustering algorithms.	Demonstrated high accuracy in predicting protein interactions. Capable of identifying binding poses within a reasonable computational time.	Moderate computational resources are required. Suitable for standard computing setups.	Predicting protein-protein interactions. Drug discovery and design.
ClusPro-AbEMap	Integrates template-based modeling with docking algorithms. Predicts epitope sites with unprecedented accuracy.	Provides accurate predictions of epitope structures and binding modes. Enhanced epitope mapping capabilities.	Requires moderate to high computational resources due to template-based modeling. Parallel processing may improve efficiency.	Accelerating discovery of potential therapeutic targets vaccine development.
RosettaDock	Employs low-resolution rigid body perturbations and high-resolution refinement. Able to explore a vast conformational space and identify low-energy structures.	Demonstrated efficacy in recovering native conformations suitable for both local and global searches.	Requires significant computational resources due to its extensive conformational sampling. Parallel processing and high-performance computing are advantageous.	Modeling antibody-antigen complexes with full backbone flexibility guiding antibody engineering efforts.
SnugDock	Tailored perturbation moves specific to antibody-binding motifs. Addresses conformational variability of antibody CDR loops.	Demonstrated superior performance in predicting antibody-antigen complexes. Particularly effective in ensemble docking approaches.	Requires moderate computational resources compared to other algorithms parallel processing may enhance performance.	Improving accuracy in docking predictions Enhancing antibody engineering and therapeutic development.

Table 2.
Comparison of the key docking algorithms, highlighting their features, accuracy, computational requirements, and applications.

partners, which may limit its applicability to cases with experimental structures available for both antibodies and antigens. Nevertheless, RosettaDock serves as a robust framework for investigating antibody-antigen interactions and guiding antibody engineering efforts.

The introduction of advanced computational platforms, such as ClusPro-AbEMap, represents a significant leap forward in epitope mapping. By integrating template-based modeling with docking algorithms, these tools can predict epitope sites with unprecedented accuracy. This approach not only accelerates the discovery of potential therapeutic targets but also opens new avenues for vaccine development and the study of immune responses.

Recent advancements include the development of SnugDock [71], a docking program tailored for antibody-antigen interactions. SnugDock optimizes paratope by CDR loop conformations and VL-VH domain orientations during docking, showcasing flexibility and success in general protein-protein docking protocols. Integration of experimental information, such as NMR data, enhances computational prediction accuracy, as demonstrated in studies like [72], where RosettaDock was applied to antibody ensemble models of PIGS and Rosetta antibody. These studies underscore the importance of considering backbone flexibility and experimental data in computational docking to improve predictive accuracy.

4.3 Advances in antibody docking

4.3.1 A stepwise docking molecular dynamics approach

The method [73] aimed to overcome the shortcomings of previous approaches by splitting the binding process into multiple steps, enabling a more realistic simulation of the docking process. Structural changes play significant roles in receptor-ligand recognition, but the underlying allosteric mechanisms are not fully addressed in the general molecular docking process. Molecular dynamics (MD) simulations have been widely used to trace the trajectory of atoms in predefined systems, aiding in the comprehension of specific biological processes.

The semi-biased stepwise docking MD was reported to yield better results, effectively recapping conformational rearrangements observed in experimental data. Traditional docking methods, on the other hand, failed to reproduce the conformational changes in the HCDR3 loop as observed in the test conditions. These findings suggest that MD-based approaches, such as the semi-biased stepwise docking MD, offer advantages over traditional docking methods in capturing complex allosteric mechanisms involved in antibody-antigen interactions.

The algorithm also highlighted important parameter considerations for the application of the stepwise docking MD method. These considerations include aspects such as binding orientation, separation distance, global influence, and endpoint criteria. Optimizing these parameters is crucial for ensuring the efficiency and accuracy of the simulation in capturing intricate allosteric mechanisms.

4.3.2 The SnugDock algorithm

SnugDock [71] represents an antibody-specific extension of the RosettaDock protocol, tailored to address the challenges associated with docking antibody structures derived from homology modeling. By incorporating antibody-specific moves and refinement steps for CDR loops, SnugDock improves docking accuracy and overcomes limitations observed in standard docking protocols. Benchmarking studies have demonstrated the superior performance of SnugDock in predicting antibody-antigen complexes, highlighting its utility in antibody engineering and therapeutic development.

Building upon the RosettaDock framework, the SnugDock algorithm represents a significant advancement in antibody-antigen docking accuracy. SnugDock introduces tailored perturbation moves specific to antibody-binding motifs, such as CDR loop remodeling and reorientation of the angle between the V domains. By addressing the conformational variability of antibody CDR loops, SnugDock enhances docking accuracy and improves the energy funnel landscape. Benchmarking studies have demonstrated the superior performance of SnugDock, particularly in ensemble docking approaches, which leverage multiple models to flatten the energy funnel and increase the proportion of high-quality structures.

4.3.3 The ClusPro program

ClusPro [74] underscores the efficacy of automated protein docking methods in accurately predicting molecular interactions, even in the absence of prior biological knowledge. A distinguishing feature of ClusPro is its utilization of structure-based potentials specifically tailored for antibody-antigen complexes. These potentials, derived from statistical mechanics principles, capture the frequency of observed interactions in protein structures, thereby enhancing the accuracy of docking predictions. By incorporating such potentials into its energy function, ClusPro achieves remarkable improvements in docking results for antibody-antigen pairs. Moreover, ClusPro's integration of the Decoys as the Reference State (DARS) potential further enhances its performance in antibody-antigen docking. By generating a diverse set of docked conformations and leveraging interaction frequency data extracted from these decoys, ClusPro effectively discriminates near-native complex structures, facilitating the identification of optimal binding orientations. Unlike traditional docking approaches that rely on symmetric pairwise interaction potentials, ClusPro's innovative methodologies, including the asymmetric DARS-type potential (ADARS), accommodate the unique characteristics of antibody-antigen interactions. By removing the constraint of interaction symmetry, ClusPro can accurately predict antibody-antigen binding orientations, offering valuable insights into molecular recognition processes.

4.3.4 Hybrid approaches for high-throughput epitope mapping

High-throughput epitope mapping remains a challenge due to the complexity of experimental techniques. Hybrid approaches that combine experimental data with computational strategies offer a promising solution to this challenge. These approaches leverage bioinformatics tools to identify interacting regions based on peptide binding assays, neutralization data, and structural analysis. By integrating diverse sources of experimental data, hybrid methods enable rapid and accurate epitope mapping, thereby facilitating the design of targeted antibody-based therapeutics.

While robust, obligate interactions are well-studied using established biochemical and structural methods and transient interactions pose challenges due to their low abundance and subtle nature. Consequently, sensitive techniques such as the yeast two-hybrid system and nuclear magnetic resonance spectroscopy are often required for their identification [75].

4.3.5 Integrating experimental restraints for enhanced docking performance

Experimental data, including alanine or site-directed mutagenesis, hydrogen-deuterium exchange mass spectrometry (HDX), NMR chemical shift perturbations, cryo-EM, and chemical crosslinking, provide valuable insights into the structural characteristics of antibody-antigen complexes [76]. By integrating these experimental restraints into docking simulations, researchers can constrain the conformational space and guide the selection of native-like docking poses. Techniques, such as RosettaDock and SnugDock, offer flexibility in incorporating experimental data, thereby enhancing the accuracy and reliability of docking predictions.

4.3.6 Artificial intelligence-driven methods to enhance predictive docking results of antibodies

A number of artificial intelligence-driven methods have emerged recently to enhance the docking accuracy of antibodies, and many such tools are currently under development (**Table 3**). PECAN, a novel AI-driven method for predicting antibody-antigen interfaces [77], employs a graph-based approach to represent antibody and antigen structures. The graphs are then fed into a neural network comprising graph convolution, attention, and fully connected layers. This network distinguishes between antibody and antigen residues within interface and non-interface regions. PECAN exhibits higher precision and recall rates compared to epitope prediction datasets [78] and paratope prediction methods [79]. Notably, these prediction methods also demonstrate high accuracies in identifying epitope and paratope regions.

AbAgIntPre-A deep learning-based prediction method [73] was created to rapidly identify antibody-antigen interactions using only amino acid sequences. The method utilizes a Siamese-like convolutional neural network architecture, which encodes the amino acid composition of both antigens and antibodies. The AbAgIntPre method revolutionizes antibody discovery by integrating the composition of k-spaced amino acid pairs (CKSAAP) encoding and a convolutional neural network (CNN) deep learning framework. This combination enables efficient prediction of antibody-antigen interactions, facilitating the identification of various antigen types. AbAgIntPre

Category	Description	Tool/software
Graph-based Neural Networks for Interface Prediction	Utilizes graph-based approach and neural network architecture to predict antibody-antigen interfaces.	PECAN shows high precision and recall rates compared to other prediction methods.
Deep Learning for Interaction Prediction	Identifies antibody-antigen interactions from amino acid sequences using convolutional neural networks.	AbAgIntPre offers generic and SARS-CoV-specific models trained on comprehensive antibody structures.
Epitope and Paratope Prediction with SVM (Support Vector Machine)	Predicts epitopes and paratopes through SVM classification of surface patches based on 3D descriptors.	EpiPred facilitates epitope prediction despite challenges in complex structure prediction.
Integrating Established Methods with Machine Learning	Combines modeling, docking, and machine learning to streamline antibody-antigen interaction characterization.	AbAdapt integrates epitope and paratope prediction with docking engines for efficient interaction analysis.

Table 3.
AI-driven methods contribute to enhancing the predictive docking results of antibodies.

offers two prediction models: a generic model and a SARS-CoV-specific model, both utilizing CNN architecture with CKSAAP sequence encoding. To enhance the data availability, comprehensive antibody structures available in the SAbDab database, along with data from other immune-related databases, the generic model was trained. This model was developed using 918 high-quality antibody-antigen complexes, which were meticulously processed and clustered into 408 subgroups to ensure the representation of diverse interactions. By assigning high-quality positive and negative antibody-antigen pairs based on this clustering, AbAgIntPre is said to have demonstrated its efficacy in predicting antigen-antibody interactions with precision and reliability.

Epitope prediction, facilitated by EpiPred, relies on geometric fitting and knowledge-based asymmetric scoring of antibody-antigen interactions [78]. This paratope prediction method makes use of a support vector machine (SVM) classifier to distinguish interface surface patches from non-interface ones based on 3D descriptors capturing global and local protein surface shapes and physicochemical properties [79]. However, it is worth noting that even recent AI-based approaches, such as AF2-multimer still face challenges in accurately predicting antibody-antigen complex structures, as previously highlighted [80].

The AbAdapt [81] tool presents an innovative approach by combining established methods for antigen/antibody modeling and docking, supplemented with machine learning at critical decision points. This server streamlines the process through automated antibody and antigen structural modeling pipelines, accommodating direct sequence input. Methods for epitope and paratope prediction were developed and trained using sequence and structural features derived from these models. Docking engines, namely Piper and Hex, were employed for global and local docking, respectively, in alignment with epitope predictions. The AbAdapt workflow involves co-clustering and re-scoring the Piper and Hex docking poses to assess the quality of top-scoring poses. Additionally, antibody-specific epitope prediction of the program integrates epitope information from antibody-antigen docking. While AbAdapt demonstrates promising results in its ability to characterize antibody-antigen interactions, areas for further improvement include pose quality and computational speed.

4.3.7 Future directions and challenges

Despite significant advancements in antibody docking algorithms and epitope mapping techniques, several challenges remain to be addressed. Improving the accuracy and scalability of docking algorithms, integrating diverse experimental data sources, and enhancing computational modeling capabilities are key areas for future research. Additionally, the development of novel bioinformatics tools and machine learning algorithms holds promise for accelerating the discovery and design of antibody-based therapeutics.

In order to enhance the precision of docking and affinity predictions for antibody-specific epitopes, substantial improvements are required. To address this need, a comprehensive benchmark dataset encompassing camelid nanobodies, therapeutic monoclonal antibodies, and broadly neutralizing antibodies targeting viral glycoproteins was assembled [82]. Another approach to enhance prediction efficacy involves exploring protein regions that resemble known antibody-binding epitopes. This method capitalizes on the ability of certain antibodies to recognize multiple antigen proteins with similar surface regions and properties [80, 83].

Acknowledgements

The authors acknowledge the use of ChatGPT for language polishing of the manuscript.

Declaration

The authors declare that no conflict of interest exists.

Author details

Shine P. Varkey[1*] and Shankar K.M.[2]

1 St. Aloysius (Deemed to Be University), Mangalore, India

2 Karnataka Veterinary, Animal and Fisheries Sciences University, Mangalore, India

*Address all correspondence to: shinepv@staloysius.ac.in.

References

[1] Van Regenmortel MH. What is a B-cell epitope? Methods in Molecular Biology (Clifton, N.J.). 2009;**524**:3-20. DOI: 10.1007/978-1-59745-450-6_1

[2] Nilvebrant J, Rockberg J. An introduction to epitope mapping. Methods in Molecular Biology (Clifton, N.J.). 2018;**1785**:1-10. DOI: 10.1007/978-1-4939-7841-0_1

[3] Desta IT, Kotelnikov S, Jones G, Ghani U, Abyzov M, Kholodov Y, et al. Mapping of antibody epitopes based on docking and homology modeling. Proteins. 2023;**91**(2):171-182. DOI: 10.1002/prot.26420

[4] Desta IT, Kotelnikov S, Jones G, Ghani U, Abyzov M, Kholodov Y, et al. The ClusPro AbEMap web server for the prediction of antibody epitopes. Nature Protocols. 2023;**18**(6):1814-1840. DOI: 10.1038/s41596-023-00826-7

[5] Salmaso V, Moro S. Bridging molecular docking to molecular dynamics in exploring ligand-protein recognition process: An overview. Frontiers in Pharmacology. 2018;**9**:923. DOI: 10.3389/fphar.2018.00923

[6] Gioia D, Bertazzo M, Recanatini M, Masetti M, Cavalli A. Dynamic docking: A paradigm shift in computational drug discovery. Molecules (Basel, Switzerland). 2017;**22**(11):2029. DOI: 10.3390/molecules22112029

[7] Sundberg EJ. Structural basis of antibody-antigen interactions. Methods in Molecular Biology (Clifton, N.J.). 2009;**524**:23-36. DOI: 10.1007/978-1-59745-450-6_2

[8] Getzoff ED, Tainer JA, Lerner RA, Geysen HM. The chemistry and mechanism of antibody binding to protein antigens. Advances in Immunology. 1988;**43**:1-98. DOI: 10.1016/s0065-2776(08)60363-6

[9] Hans D, Young PR, Fairlie DP. Current status of short synthetic peptides as vaccines. Medicinal Chemistry (Shariqah (United Arab Emirates)). 2006;**2**(6):627-646. DOI: 10.2174/157340641060206062

[10] Toride King M, Brooks CL. Epitope mapping of antibody-antigen interactions with X-ray crystallography. Methods in Molecular Biology (Clifton, N.J.). 2018;**1785**:13-27. DOI: 10.1007/978-1-4939-7841-0_2

[11] Schnur E, Kessler N, Zherdev Y, Noah E, Scherf T, Ding FX, et al. NMR mapping of RANTES surfaces interacting with CCR5 using linked extracellular domains. The FEBS Journal. 2013;**280**(9):2068-2084. DOI: 10.1111/febs.12230

[12] Säfsten P. Epitope mapping by surface plasmon resonance. Methods in Molecular Biology (Clifton, N.J.). 2009;**524**:67-76. DOI: 10.1007/978-1-59745-450-6_5

[13] Ramaraj T, Angel T, Dratz EA, Jesaitis AJ, Mumey B. Antigen-antibody interface properties: Composition, residue interactions, and features of 53 non-redundant structures. Biochimica et Biophysica Acta. 2012;**1824**(3):520-532. DOI: 10.1016/j.bbapap.2011.12.007

[14] Wu D, Sun J, Xu T, Wang S, Li G, Li Y, et al. Stacking and energetic contribution of aromatic islands at the binding interface of antibody proteins. Immunome Research. 2010;**6**(Suppl. 1(Suppl. 1)):S1. DOI: 10.1186/1745-7580-6-S1-S1

[15] Shiroishi M, Tsumoto K, Tanaka Y, Yokota A, Nakanishi T, Kondo H, et al. Structural consequences of mutations in interfacial Tyr residues of a protein antigen-antibody complex: The case of HyHEL-10-HEL. The Journal of Biological Chemistry. 2007;**282**(9):6783-6791. DOI: 10.1074/jbc.M605197200

[16] Madsen AV, Mejias-Gomez O, Pedersen LE, Morth JP, Kristensen P, Jenkins TP, et al. Structural trends in antibody-antigen binding interfaces: A computational analysis of 1833 experimentally determined 3D structures. Computational and Structural Biotechnology Journal. 2023;**23**:199-211. DOI: 10.1016/j.csbj.2023.11.056

[17] Sundberg EJ, Mariuzza RA. Molecular recognition in antibody-antigen complexes. Advances in Protein Chemistry. 2002;**61**:119-160. DOI: 10.1016/s0065-3233(02)61004-6

[18] Davies DR, Cohen GH. Interactions of protein antigens with antibodies. Proceedings of the National Academy of Sciences of the United States of America. 1996;**93**(1):7-12. DOI: 10.1073/pnas.93.1.7

[19] Malby RL, Tulip WR, Harley VR, McKimm-Breschkin JL, Laver WG, Webster RG, et al. The structure of a complex between the NC10 antibody and influenza virus neuraminidase and comparison with the overlapping binding site of the NC41 antibody. Structure. 1994;**2**(8):733-746

[20] Tomlinson IM, Walter G, Jones PT, Dear PH, Sonnhammer EL, Winter G. The imprint of somatic hypermutation on the repertoire of human germline V genes. Journal of Molecular Biology. 1996;**256**(5):813-817. DOI: 10.1006/jmbi.1996.0127

[21] Tulip WR, Harley VR, Webster RG, Novotny J. N9 neuraminidase complexes with antibodies NC41 and NC10: Empirical free energy calculations capture specificity trends observed with mutant binding data. Biochemistry. 1994;**33**(26):7986-7997. DOI: 10.1021/bi00192a002

[22] Jin L, Wells JA. Dissecting the energetics of an antibody-antigen interface by alanine shaving and molecular grafting. Protein Science: A Publication of the Protein Society. 1994;**3**(12):2351-2357. DOI: 10.1002/pro.5560031219

[23] Novotny J, Sharp K. Electrostatic fields in antibodies and antibody/antigen complexes. Progress in Biophysics and Molecular Biology. 1992;**58**(3):203-224. DOI: 10.1016/0079-6107(92)90006-r

[24] Cunningham BC, Wells JA. Comparison of a structural and a functional epitope. Journal of Molecular Biology. 1993;**234**(3):554-563. DOI: 10.1006/jmbi.1993.1611

[25] Fersht AR, Shi JP, Knill-Jones J, Lowe DM, Wilkinson AJ, Blow DM, et al. Hydrogen bonding and biological specificity analysed by protein engineering. Nature. 1985;**314**(6008):235-238. DOI: 10.1038/314235a0

[26] Murphy KP, Gill SJ. Solid model compounds and the thermodynamics of protein unfolding. Journal of Molecular Biology. 1991;**222**(3):699-709. DOI: 10.1016/0022-2836(91)90506-2

[27] Pace CN. Contribution of the hydrophobic effect to globular protein stability. Journal of Molecular Biology. 1992;**226**(1):29-35. DOI: 10.1016/0022-2836(92)90121-y

[28] Ysern X, Li H, Mariuzza RA. Imperfect interfaces. Nature Structural Biology. 1998;**5**(6):412-414. DOI: 10.1038/nsb0698-412

[29] Yokota A, Tsumoto K, Shiroishi M, Kondo H, Kumagai I. The role of hydrogen bonding via interfacial water molecules in antigen-antibody complexation: The HyHEL-10-HEL interaction. The Journal of Biological Chemistry. 2003;**278**(7):5410-5418. DOI: 10.1074/jbc.M210182200

[30] Jen-Jacobson L, Engler LE, Jacobson LA. Structural and thermodynamic strategies for site-specific DNA binding proteins. Structure (London, England: 1993). 2000;**8**(10):1015-1023. DOI: 10.1016/s0969-2126(00)00501-3

[31] Calderone CT, Williams DH. An enthalpic component in cooperativity: The relationship between enthalpy, entropy, and noncovalent structure in weak associations. Journal of the American Chemical Society. 2001;**123**(26):6262-6267. DOI: 10.1021/ja003016y

[32] Borea PA, Dalpiaz A, Varani K, Gilli P, Gilli G. Can thermodynamic measurements of receptor binding yield information on drug affinity and efficacy? Biochemical Pharmacology. 2000;**60**(11):1549-1556. DOI: 10.1016/s0006-2952(00)00368-3

[33] Braden BC, Poljak RJ. Structural features of the reactions between antibodies and protein antigens. FASEB Journal : Official Publication of the Federation of American Societies for Experimental Biology. 1995;**9**(1):9-16. DOI: 10.1096/fasebj.9.1.7821765

[34] Ladbury JE, Wright JG, Sturtevant JM, Sigler PB. A thermodynamic study of the trp repressor-operator interaction. Journal of Molecular Biology. 1994;**238**(5):669-681. DOI: 10.1006/jmbi.1994.1328

[35] Sleigh SH, Seavers PR, Wilkinson AJ, Ladbury JE, Tame JR. Crystallographic and calorimetric analysis of peptide binding to OppA protein. Journal of Molecular Biology. 1999;**291**(2):393-415. DOI: 10.1006/jmbi.1999.2929

[36] Holdgate GA, Tunnicliffe A, Ward WH, Weston SA, Rosenbrock G, Barth PT, et al. The entropic penalty of ordered water accounts for weaker binding of the antibiotic novobiocin to a resistant mutant of DNA gyrase: A thermodynamic and crystallographic study. Biochemistry. 1997;**36**(32):9663-9673. DOI: 10.1021/bi970294

[37] Lawrence MC, Colman PM. Shape complementarity at protein/protein interfaces. Journal of Molecular Biology. 1993;**234**(4):946-950. DOI: 10.1006/jmbi.1993.1648

[38] Cohen GH, Silverton EW, Padlan EA, Dyda F, Wibbenmeyer JA, Willson RC, et al. Water molecules in the antibody-antigen interface of the structure of the fab HyHEL-5-lysozyme complex at 1.7 A resolution: Comparison with results from isothermal titration calorimetry. Acta Crystallographica. Section D, Biological Crystallography. 2005;**61**(Pt 5):628-633. DOI: 10.1107/S0907444905007870

[39] Kuroda D, Gray JJ. Shape complementarity and hydrogen bond preferences in protein-protein interfaces: Implications for antibody modeling and protein-protein docking. Bioinformatics (Oxford, England). 2016;**32**(16):2451-2456. DOI: 10.1093/bioinformatics/btw197

[40] Tsumoto K, Caaveiro JM. Antigen–antibody binding. In: eLS. John Wiley & Sons, Ltd; 2016. DOI: 10.1002/9780470015902.a0001117.pub3

[41] Al-Lazikani B, Lesk AM, Chothia C. Standard conformations for the canonical structures of immunoglobulins. Journal

of Molecular Biology. 1997;**273**(4):927-948. DOI: 10.1006/jmbi.1997.1354

[42] Morea V, Lesk AM, Tramontano A. Antibody modeling: Implications for engineering and design. Methods (San Diego, Calif.). 2000;**20**(3):267-279. DOI: 10.1006/meth.1999.0921

[43] Collis AV, Brouwer AP, Martin AC. Analysis of the antigen combining site: Correlations between length and sequence composition of the hypervariable loops and the nature of the antigen. Journal of Molecular Biology. 2003;**325**(2):337-354. DOI: 10.1016/s0022-2836(02)01222-6

[44] Bogan AA, Thorn KS. Anatomy of hot spots in protein interfaces. Journal of Molecular Biology. 1998;**280**(1):1-9. DOI: 10.1006/jmbi.1998.1843

[45] Raghunathan G, Smart J, Williams J, Almagro JC. Antigen-binding site anatomy and somatic mutations in antibodies that recognize different types of antigens. Journal of Molecular Recognition: JMR. 2012;**25**(3):103-113. DOI: 10.1002/jmr.2158

[46] Chothia C, Lesk AM, Tramontano A, Levitt M, Smith-Gill SJ, Air G, et al. Conformations of immunoglobulin hypervariable regions. Nature. 1989;**342**(6252):877-883. DOI: 10.1038/342877a0

[47] Stanfield RL, Wilson IA. Antigen-induced conformational changes in antibodies: A problem for structural prediction and design. Trends in Biotechnology. 1994;**12**(7):275-279. DOI: 10.1016/0167-7799(94)90139-2

[48] North B, Lehmann A, Dunbrack RL Jr. A new clustering of antibody CDR loop conformations. Journal of Molecular Biology. 2011;**406**(2):228-256. DOI: 10.1016/j.jmb.2010.10.030

[49] Clark LA, Ganesan S, Papp S, van Vlijmen HW. Trends in antibody sequence changes during the somatic hypermutation process. Journal of Immunology (Baltimore, Md. : 1950). 2006;**177**(1):333-340. DOI: 10.4049/jimmunol.177.1.333

[50] Brooks CL, Blackler RJ, Sixta G, Kosma P, Müller-Loennies S, Brade L, et al. The role of CDR H3 in antibody recognition of a synthetic analog of a lipopolysaccharide antigen. Glycobiology. 2010;**20**(2):138-147. DOI: 10.1093/glycob/cwp150

[51] Julien JP, Huarte N, Maeso R, Taneva SG, Cunningham A, Nieva JL, et al. Ablation of the complementarity-determining region H3 apex of the anti-HIV-1 broadly neutralizing antibody 2F5 abrogates neutralizing capacity without affecting core epitope binding. Journal of Virology. 2010;**84**(9):4136-4147. DOI: 10.1128/JVI.02357-09

[52] Ofek G, Guenaga FJ, Schief WR, Skinner J, Baker D, Wyatt R, et al. Elicitation of structure-specific antibodies by epitope scaffolds. Proceedings of the National Academy of Sciences of the United States of America. 2010;**107**(42):17880-17887. DOI: 10.1073/pnas.1004728107

[53] Kuroda D, Shirai H, Kobori M, Nakamura H. Structural classification of CDR-H3 revisited: A lesson in antibody modeling. Proteins. 2008;**73**(3):608-620. DOI: 10.1002/prot.22087

[54] Soto CS, Fasnacht M, Zhu J, Forrest L, Honig B. Loop modeling: Sampling, filtering, and scoring. Proteins. 2008;**70**(3):834-843. DOI: 10.1002/prot.21612

[55] Sivasubramanian A, Sircar A, Chaudhury S, Gray JJ. Toward high-resolution homology modeling of

antibody Fv regions and application to antibody-antigen docking. Proteins. 2009;**74**(2):497-514. DOI: 10.1002/prot.22309

[56] Gray JJ, Moughon S, Wang C, Schueler-Furman O, Kuhlman B, Rohl CA, et al. Protein-protein docking with simultaneous optimization of rigid-body displacement and side-chain conformations. Journal of Molecular Biology. 2003;**331**(1):281-299. DOI: 10.1016/s0022-2836(03)00670-3

[57] Wang C, Schueler-Furman O, Andre I, London N, Fleishman SJ, Bradley P, et al. RosettaDock in CAPRI rounds 6-12. Proteins. 2007;**69**(4):758-763. DOI: 10.1002/prot.21684

[58] Rapp CS, Pollack RM. Crystal packing effects on protein loops. Proteins. 2005;**60**(1):103-109. DOI: 10.1002/prot.20492

[59] Sircar A, Kim ET, Gray JJ. RosettaAntibody: Antibody variable region homology modeling server. Nucleic Acids Research. 2009;**37**(Web Server issue):W474-W479. DOI: 10.1093/nar/gkp387

[60] Whitelegg NR, Rees AR. WAM: An improved algorithm for modelling antibodies on the WEB. Protein Engineering. 2000;**13**(12):819-824. DOI: 10.1093/protein/13.12.819

[61] Chailyan A, Marcatili P, Tramontano A. The association of heavy and light chain variable domains in antibodies: Implications for antigen specificity. The FEBS Journal. 2011;**278**(16):2858-2866. DOI: 10.1111/j.1742-4658.2011.08207.x

[62] Abhinandan KR, Martin AC. Analysis and improvements to Kabat and structurally correct numbering of antibody variable domains. Molecular Immunology. 2008;**45**(14):3832-3839. DOI: 10.1016/j.molimm.2008.05.022

[63] Kuroda D, Shirai H, Jacobson MP, Nakamura H. Computer-aided antibody design. Protein Engineering, Design and Selection: PEDS. 2012;**25**(10):507-521. DOI: 10.1093/protein/gzs024

[64] Wang N, Smith WF, Miller BR, Aivazian D, Lugovskoy AA, Reff ME, et al. Conserved amino acid networks involved in antibody variable domain interactions. Proteins. 2009;**76**(1):99-114. DOI: 10.1002/prot.22319

[65] Sivasubramanian A, Chao G, Pressler HM, Wittrup KD, Gray JJ. Structural model of the mAb 806-EGFR complex using computational docking followed by computational and experimental mutagenesis. Structure (London, England: 1993). 2006;**14**(3):401-414. DOI: 10.1016/j.str.2005.11.022

[66] Gray JJ. High-resolution protein-protein docking. Current Opinion in Structural Biology. 2006;**16**(2):183-193. DOI: 10.1016/j.sbi.2006.03.003

[67] Zacharias M. Accounting for conformational changes during protein-protein docking. Current Opinion in Structural Biology. 2010;**20**(2):180-186. DOI: 10.1016/j.sbi.2010.02.001

[68] Bonvin AM. Flexible protein-protein docking. Current Opinion in Structural Biology. 2006;**16**(2):194-200. DOI: 10.1016/j.sbi.2006.02.002

[69] Chaudhury S, Gray JJ. Conformer selection and induced fit in flexible backbone protein-protein docking using computational and NMR ensembles. Journal of Molecular Biology. 2008;**381**(4):1068-1087. DOI: 10.1016/j.jmb.2008.05.042

[70] Lyskov S, Gray JJ. The RosettaDock server for local protein-protein docking. Nucleic Acids Research. 2008;**36**(Web Server issue):W233-W238. DOI: 10.1093/nar/gkn216

[71] Sircar A, Gray JJ. SnugDock: Paratope structural optimization during antibody-antigen docking compensates for errors in antibody homology models. PLoS Computational Biology. 2010;**6**(1):e1000644. DOI: 10.1371/journal.pcbi.1000644

[72] Simonelli L, Beltramello M, Yudina Z, Macagno A, Calzolai L, Varani L. Rapid structural characterization of human antibody-antigen complexes through experimentally validated computational docking. Journal of Molecular Biology. 2010;**396**(5):1491-1507. DOI: 10.1016/j.jmb.2009.12.053

[73] Huang Y, Li Z, Hong Q, Zhou L, Ma Y, Hu Y, et al. A stepwise docking molecular dynamics approach for simulating antibody recognition with substantial conformational changes. Computational and Structural Biotechnology Journal. 2022;**20**:710-720. DOI: 10.1016/j.csbj.2022.01.012

[74] Brenke R, Hall DR, Chuang GY, Comeau SR, Bohnuud T, Beglov D, et al. Application of asymmetric statistical potentials to antibody-protein docking. Bioinformatics (Oxford, England). 2012;**28**(20):2608-2614. DOI: 10.1093/bioinformatics/bts493

[75] Peng HP, Lee KH, Jian JW, Yang AS. Origins of specificity and affinity in antibody-protein interactions. Proceedings of the National Academy of Sciences of the United States of America. 2014;**111**(26):E2656-E2665. DOI: 10.1073/pnas.1401131111

[76] Tran MH, Schoeder CT, Schey KL, Meiler J. Computational structure prediction for antibody-antigen complexes from hydrogen-deuterium exchange mass spectrometry: Challenges and outlook. Frontiers in Immunology. 2022;**13**:859964. DOI: 10.3389/fimmu.2022.859964

[77] Pittala S, Bailey-Kellogg C. Learning context-aware structural representations to predict antigen and antibody binding interfaces. Bioinformatics (Oxford, England). 2020;**36**(13):3996-4003. DOI: 10.1093/bioinformatics/btaa263

[78] Krawczyk K, Liu X, Baker T, Shi J, Deane CM. Improving B-cell epitope prediction and its application to global antibody-antigen docking. Bioinformatics. 2014;**30**:2288-2294. DOI: 10.1093/bioinformatics/btu190

[79] Daberdaku S, Ferrari C. Antibody interface prediction with 3D Zernike descriptors and SVM. Bioinformatics. 2019;**35**:1870-1876. DOI: 10.1093/bioinformatics/bty918

[80] Tsuchiya Y, Yamamori Y, Tomii K. Protein–protein interaction prediction methods: From docking-based to AI-based approaches. Biophysical Reviews. 2022;**14**:1341-1348. DOI: 10.1007/s12551-022-01032-7

[81] Davila A, Xu Z, Li S, Rozewicki J, Wilamowski J, Kotelnikov S, et al. AbAdapt: An adaptive approach to predicting antibody-antigen complex structures from sequence. Bioinformatics Advances. 2022;**2**(1):vbac015. DOI: 10.1093/bioadv/vbac015

[82] Guest JD, Vreven T, Zhou J, Moal I, Jeliazkov JR, Gray JJ, et al. An expanded benchmark for antibody-antigen docking and affinity prediction reveals insights into antibody recognition determinants. Structure. 2021;**29**:606-621.e5. DOI: 10.1016/J.STR.2021.01.005

[83] Negi SS, Braun W. Cross-react: A new structural bioinformatics method for predicting allergen cross-reactivity. Bioinformatics (Oxford, England). 2017;**33**(7):1014-1020. DOI: 10.1093/bioinformatics/btw767

Section 2

Applications

Chapter 5

Applications of Molecular Docking Studies in SARS-CoV-2 Targeted Drug Discovery and the Gains Achieved through Molecular Docking

Merve Yildirim and Ismail Celik

Abstract

In this chapter, we delve into the pivotal role of molecular docking in the realm of computational biology and chemistry, focusing specifically on its application in drug discovery targeting SARS-CoV-2. Molecular docking, a critical computational technique, has played a significant role in predicting the interactions and bindings of molecules, particularly concerning SARS-CoV-2's main protease and RNA polymerase. This chapter highlights the synergy between molecular docking and virtual screening, emphasizing the expedited identification and evaluation of potential drug candidates against SARS-CoV-2. Through a comprehensive discussion, we aim to provide a nuanced understanding of the rapid advancements in drug discovery for SARS-CoV-2, accentuating the indispensable value of computational tools and methods in contemporary therapeutic development.

Keywords: SARS-CoV-2, main protease, RNA polymerase, virtual screening, molecular docking

1. Introduction

Severe acute respiratory syndrome coronavirus (SARS-CoV) and Middle East respiratory syndrome coronavirus (MERS-CoV) are viruses in the family Coronaviridae that infect different animals and can cause mild or severe respiratory infections in humans. These viruses were detected in humans in 2002 and 2012, with fatal outcomes. In late 2019, a new coronavirus emerged in Wuhan, China. The International Committee on Taxonomy of Viruses named it SARS-CoV-2, while the WHO named the disease COVID-19. This virus multiplied rapidly in a short time and spread all over the world [1]. As of October 14, 2023, the World Health Organization reported a cumulative total of 774,291,287 COVID-19 cases and 7,019,704 deaths worldwide. Vaccination began on June 22, 2020, and a total of 13.59

IntechOpen

billion vaccine doses have been administered "https://.covid19.who.int/ (accessed on 29 January 2024)".

Coronaviruses are enveloped, positive-sense single-stranded RNA viruses with genome sizes ranging between 26 and 32 kilobases. The genomic structure comprises various components that form the enveloped virion. The genome of SARS-CoV-2 shows approximately 80% similarity to that of SARS-CoV. The virion consists of genomic RNA, phosphorylated nucleocapsid (N) protein, membrane (M) protein, envelope (E) protein, and spike glycoprotein (S) as fundamental building blocks (**Figure 1**). The life cycle of SARS-CoV-2 begins with the binding of the viral protein S to the angiotensin-converting enzyme-2 (ACE2) receptor [2–6].

Although the pathogenesis of SARS-CoV-2 infection remains uncertain, severe COVID-19 is characterized by an uncontrolled immune response induced by the host's reaction to the virus. This leads to a sudden and explosive release of various immune mediators, particularly cytokines, and damage-associated molecular patterns (DAMPs). The unregulated immune response can result in multiple organ dysfunction, leading to sepsis or septic shock, and ultimately giving rise to a life-threatening syndrome [4]. Numerous studies have been conducted on viruses within the Coronaviridae family. It has been revealed that the ACE-2 receptor plays a crucial role in the virus's entry into the host. Additionally, RNA-dependent RNA polymerase (RdRp), main protease (Mpro), helicase, papain-like protease, and mTOR signaling pathway [7] in the virus are identified as potential targets [8].

The COVID-19 virus can cause a range of symptoms, including a dry cough, high fever, body pain, and shortness of breath. These symptoms may appear within 2–14 days of exposure to the virus. In more severe cases, patients may develop pneumonia, as well as neurological symptoms such as headache, loss of taste and smell, and visual impairment. It is important to seek medical attention if you experience any of these symptoms [6, 9].

Since the commencement of the pandemic in 2019, the virus has undergone numerous changes, giving rise to the development of novel variants and their subvariants. Each variant carries unique genetic markers that may impact its transmissibility, virulence, and ability to evade the body's immune system [10]. The fact that there are many different variants of the virus, in addition to the well-known ones such as Alpha, Beta, Gamma, Omicron, and Pirola, highlights the urgent situation we are facing. This is evident from the high rates of transmission and mortality. As a result, it

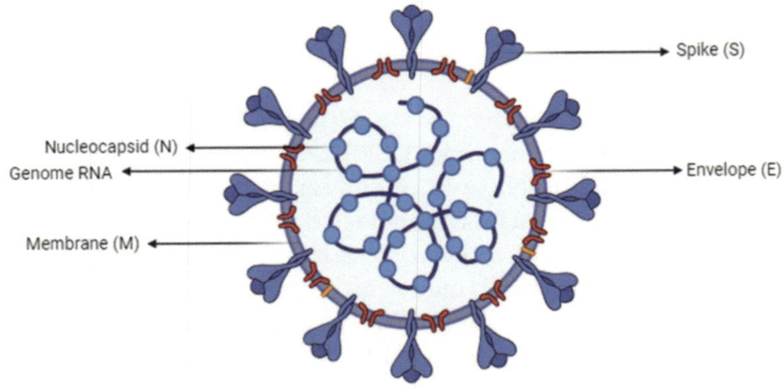

Figure 1.
The schematic structure of the SARS-CoV-2 virus created in BioRender.com (accessed on January 31, 2024).

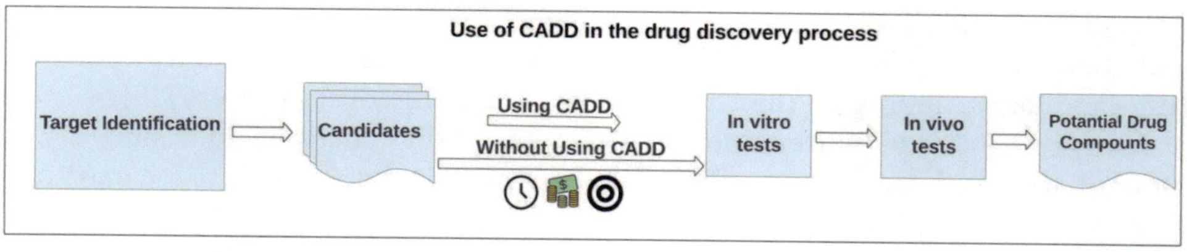

Figure 2.
The role of computer-aided drug design in the drug discovery process.

is crucial to identify effective drug targets, antiviral drug molecules, and strategies to contain the spread of the virus quickly and accurately [11].

Due to the limited number of drugs approved to treat COVID-19, it has become a significant threat to global health, safety, and the economy. The urgency to find antiviral agents to combat the spread of the SARS-CoV virus has made it necessary to expedite drug discovery and development. Researchers have had to employ new strategies to hasten traditional processes and discover new drug candidates more quickly. One of the most crucial solutions has been the use of computer-aided drug design (CADD). CADD is a computational approach used to discover, develop, and analyze drugs and active molecules with similar biochemical properties. It can be applied to all stages of drug development, making it essential because of its high prediction accuracies and unique conveniences in terms of time and resources. As a result, the rapid advancements in drug discovery during this period increased the scientific community's capacity to provide faster and more effective solutions to urgent health issues [12–14].

Once the target is determined, the process of selecting candidates for in vitro testing begins. This step is slowly progressing without the aid of computer-aided drug design, also known as in silico studies. However, by applying in silico studies, it is possible to select the compounds most suitable for the target structure and eliminate the unsuitable compounds. This approach helps speed up the process and achieve a higher success rate more cost-effectively. At the end of in vitro and in vivo experiments, a potential drug molecule is obtained (**Figure 2**) [15].

2. Molecular docking in computational biology

Molecular docking is a computational tool that plays a crucial role in computer-aided drug design. It is a molecular mechanic approach that aims to fit a ligand into a three-dimensional binding site. It helps in identifying potential drug candidates and studying the interactions between molecules at the molecular level, which is essential for drug design. This technique can predict the conformation of molecular interactions and calculate the binding energy scores of these interactions. It simulates the process of evaluating how efficiently a drug candidate can bind to a target molecule. Molecular docking has become a routine part of computer-aided drug design and plays a vital role in drug discovery [16–18].

With advances in the field of drug discovery, the use of in silico tools has become significantly widespread today. Molecular modeling contributes significantly to drug design processes, especially by providing a critical tool to understand the fundamental interactions of ligands and the mode of action of the target molecule. Methods such as molecular docking are effective in identifying different modes of docking through key docking scores and binding interactions. These in silico assessments contribute to the rapid and cost-effective identification and optimization of potential drug candidates, improving the efficiency of the drug discovery process [19].

Many research groups are exploring drug repurposing, focusing on in silico evaluations of approved drugs. Drug repurposing is a process that involves using existing drugs for new indications. Compared to traditional drug development processes, drug repurposing offers several advantages. Firstly, it facilitates a faster development timeline due to the already proven safety of drugs used in humans. By bypassing early clinical trial stages, the timeline is shortened, and costs are reduced. Secondly, utilizing approved drugs for new indications accelerates the creation of new treatments, decreasing the costs of drug discovery and development while enhancing the potential for rapid access to innovative therapies. However, there are challenges and limitations associated with drug repurposing [20–22]. Data from 2020 showed a significant increase in virtual screening studies to find new potent molecules or repurpose approved, withdrawn, or orphan drugs effective against the virus [13].

There are different types of software available to perform molecular docking, which include AutoDock [23], AutoDock Vina [24], DOCK, FlexX [25], Molecular Operating Environment (MOE) [26], and Glide [27, 28]. These software programs have been used in various molecular docking studies such as Flexx and AutoDock Vina for Mpro [29, 30], AutoDock for proteins [31, 32], MOE for ACE-2 [33], and Glide for Rbrd [34], which are the target structures of SARS-CoV-2, with promising results.

The process of molecular docking involves evaluating the quality of the pose using a scoring function, along with predicting the position and conformation of the ligand in the protein binding site. Another crucial task in docking is scoring active compounds higher than known inactive ones, but this can be difficult due to various factors. The sampling process has many degrees of freedom, such as the rotation and translation of one molecule relative to another. Therefore, efficiently sampling conformational space is still a challenge in molecular docking. Early approaches reduced the degrees of freedom by treating the ligand and protein as rigid bodies, but those relied on shape similarities between the ligand and the protein binding site.

Modeling molecular flexibility poses challenges in terms of computational efficiency, as conformation is linked to protein-ligand interactions. As a result, many docking programs now consider the entire conformational space of the ligand. Thus, molecular docking algorithms aim to sample the conformational space of the ligand and protein and evaluate their interactions. However, they can face challenges like flexibility, computational efficiency, and scoring accuracy [35].

Previous studies have identified three types of scoring functions for protein-ligand interactions: physics-based, empirical, and knowledge-based. However, these scoring functions have limitations, and no single method can perform perfectly in all aspects. To improve our understanding of scoring functions, a new classification scheme has been proposed. This new scheme categorizes scoring functions into four groups: physics-based, empirical, knowledge-based, and machine learning-based. Hybrid scoring functions can improve performance by combining different scoring functions. These scoring functions can compensate for the weaknesses of different scoring functions. However, the computation time of hybrid scoring functions may be longer.

Machine learning-based scoring functions can outperform traditional scoring functions and improve prediction accuracy by using extended training datasets. These scoring functions may be more widely used in the future. Combining different types of scoring functions and using new types of features can improve the performance of scoring functions in molecular docking. Additionally, these scoring functions are open source [36].

The optimization of the scoring function is done by using large protein-ligand complex structures that have been experimentally determined. In recent years, the development of technologies such as infrared spectroscopy, X-ray crystallography, nuclear magnetic resonance spectroscopy, and cryogenic electron microscopy has increased the possibilities for protein-ligand complex structure determination. This has led to efforts to optimize scoring functions and improve sampling procedures. One example is the Lamarck genetic algorithm used in AutoDock Vina, which is implemented in an OpenMP-based multithreaded version to search for ligand conformations. There are also efforts to improve the performance of these algorithms by using Graphics Processing Units (GPU)-based accelerations [37].

The COVID-19 pandemic has highlighted the importance of using computer simulations to screen small molecules for potential viral inhibitors. Despite the promising results of molecular docking studies, receptor flexibility, especially the movement of the backbone and other important secondary elements, can make molecular docking difficult. Nevertheless, successful applications have demonstrated the efficiency of computational approaches for screening large databases and designing new molecules. To further advance molecular docking, the development of accurate and cost-effective scoring functions is crucial. However, it is important to note that experimental technology still plays a critical role in producing realistic molecular interactions [16, 38–40]. After conducting molecular docking studies, experimental studies are carried out to further analyze the structural and spectral properties of the molecules. Through numerous studies, the computational analysis of various properties is compared to experimental findings. This process leads to the confirmation of the molecular docking data by comparing it to the experimental results [41, 42].

3. SARS-CoV-2 targeted drug discovery

Ongoing research is being conducted to develop antiviral agents that can be effective against coronaviruses. These studies have identified several potential drug targets, including the human receptor ACE2, which facilitates the entry of the virus into the host, the spike protein of the virus, RdRp, Mpro, and papain-like protease [8, 11, 43].

Although SARS-CoV-2 interacts with all drug targets, its interaction with the active site residues of the Mpro was found to be much stronger than with the other targets, spike and ACE2. On the other hand, the spike protein has been used in vaccine studies, but frequent mutations in this protein have been an obstacle. The mutation observed in the spike protein was not observed in the Mpro. These reasons have put the Mpro several steps ahead of other targets in drug discovery and use as a drug target [44–46].

There are approved drugs with proven efficacy on the Mpro and RdRp among the targeted structures. Mpro and RdRp 3D structures from the protein data bank [https://www.rcsb.org/ accessed on: 27.01.2024] were visualized in the UCSF Chimera v1.17.3 program (**Figure 3**). The virus requires the enzyme Mpro for proteolytic processing of polyproteins. The active site of Mpro is primarily characterized by a catalytic dyad composed of His41 and Cys145. These residues are crucial for the enzymatic activity, as they facilitate the hydrolysis of peptide bonds within the polyprotein. Inhibitors designed to target this active site aim to disrupt the proteolytic process,

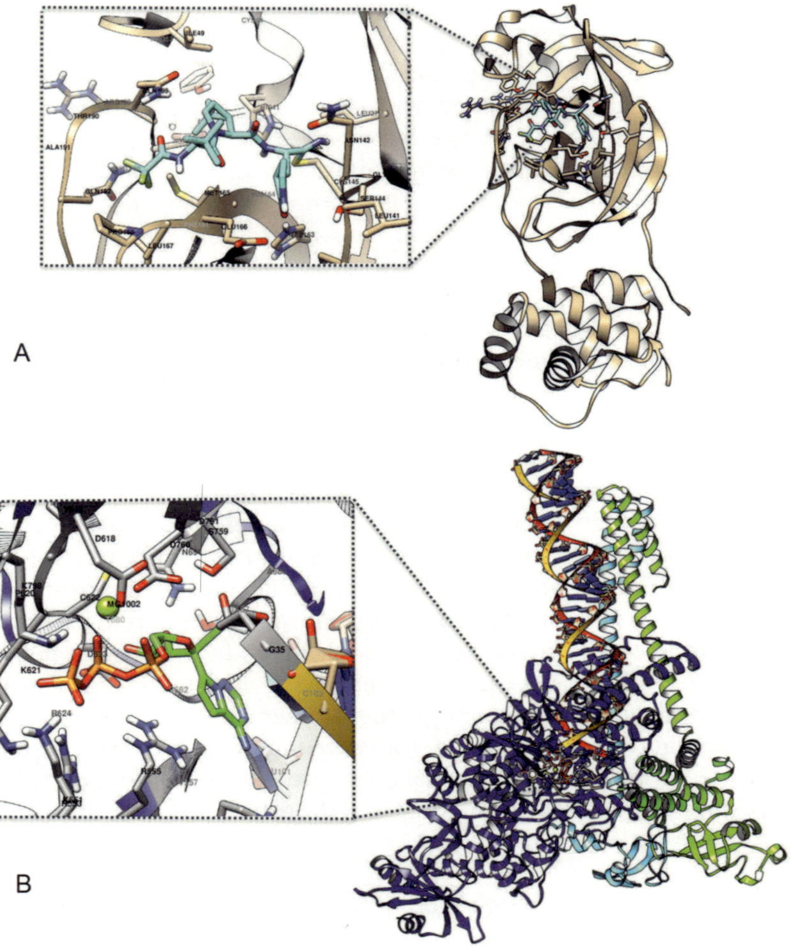

Figure 3.
A: 3D structure and protein-ligand interactions of SARS-CoV-2 main protease (Mpro) with Nirmatrelvir (PDB:8E25) B: RNA-dependent RNA polymerase (RdRp) with Remdesivir triphosphate (PDB: 7UO4) [47, 48].

thereby impeding viral replication. Mpro inhibitors inhibit viral replication and transcription of the virus by binding to the active site of the protease (**Figure 3A**). Rupintrivir, Lopinavir/Ritonavir, Nirmatrelvir, and Paxlovid [nirmatrelvir + ritonavir] are examples of this group of drugs [2, 3, 8, 14, 39, 49, 50].

RdRp is a critical viral enzyme that catalyzes RNA replication from an RNA template and mediates viral replication and transcription. The catalytic activity of RdRp involves the synthesis of complementary RNA chains for a specific RNA template, presenting potential therapeutic targets against viral infections, as this activity is not essential for the survival of eukaryotic cells [51]. The active site of the RNA synthesis process comprises cofactor Mg2+ and several motifs (A to G), with motifs A and C being particularly important. These motifs include residues Asp618, Asp760, Asp761, Asp762 (motif A), and Ser861, Lys545, and Arg555 (motif C), which play a central role in catalysis (**Figure 3B**). Nucleotide analogs designed to inhibit the replication process of the virus target the mechanism of nucleotide addition at this active site. Remdesivir and Molnupiravir in this group inhibit replication by stopping the development of the RNA chain [14, 52].

A significant number of new variants and subtypes of the virus have emerged rapidly, each carrying unique genetic signatures that may influence their contagiousness, virulence, and the potential to evade the immune system. Alongside this diversity, challenges in in silico studies include inadequate resolutions of crystallographic targets, both structural and conformational flexibility issues, as well as simplifications and assumptions that adversely affect the accuracy of scoring functions. Additionally, limitations in properly accounting for solvent and entropic effects, hydrogen bonding, directional interactions, etc., can lead to difficulties [10, 13].

Most molecular docking software programs rely on force field calculations driven by quantum mechanics [QM] and experimental data to predict binding energy. However, accurate binding energies can only be determined through methods such as density functional theory [DFT] and molecular dynamics simulations. Therefore, it is crucial to employ multiple methods, rather than relying solely on the in silico approach [13].

A study titled "An oral SARS-CoV-2 Mpro inhibitor clinical candidate for the treatment of COVID-19" was published in 2021. It discusses the discovery of the oral drug PF-07321332/ Nirmatrevir, which was developed as a Mpro inhibitor to treat COVID-19. The drug has reached Phase I stages and has been proven effective against all coronaviruses. The study provides information on the bonds formed between the target and the ligands, bond lengths, type, and degree of interaction. For example, the study showed that the number of hydrogen bond donors was linked with poor oral bioavailability. Therefore, nitriles and benzothiazol-2-yl ketones were considered. When the nitrile compound was used, there was an increase in oral absorption in rats, indicating that the interaction could be improved by taking advantage of the structural features. The study also considered the structural cavities of the protease. Introducing suitable compounds into these pockets creates hydrogen bonds and an interaction. The drug PF-07321332/Nirmatrevir has been identified as a potent inhibitor of SARS-CoV-2 Mpro biochemical activity with its nitrile group [52, 53].

We observe not only in this study but in many studies that target-ligand complexes achieve results by taking advantage of interactions. These findings highlight the importance of molecular docking in the drug discovery process. In a study examined against the Omicron variant, it was explained that different types of interactions between ligand atoms and protein residues determine the extent of inhibition of the target protein. In this research, in silico studies, molecular docking and molecular dynamics were used. The number of hydrogen bonds in the docking results of the compounds found to be active was considered the most important of these interactions, as it ensures target-ligand stability. Additionally, other types of interactions such as pi-alkyl, pi-cation, and pi-sigma were also observed in the active compounds. The importance of these interactions in the activity is emphasized [12, 54, 55].

4. Integration of molecular docking and virtual screening

Virtual screening, considered a computational method proven and appreciated, serves as a computer-aided complement to experimental high-throughput screening, which is a contemporary of hit identification and optimization. It can be classified into two main categories: ligand-based and structure-based methods. Ligand-based methods, such as pharmacophore modeling and quantitative structure-activity relationship (QSAR) techniques, can be employed when there is minimal or no structural information available for the targets, and a series of known active ligand molecules exists. On the other hand, molecular docking stands out as a widely applied method in structure-based drug design since the early 1980s. Approved in 1981, Captopril (Capoten) was developed using structure-activity relationship (SAR), structure-based drug design (SBDD), and ligand-based drug design (LBDD). The use of CADD has continued to increase since then [13, 56]. Various programs based on different algorithms have been developed to conduct molecular docking studies, contributing to its growing significance as a valuable tool in pharmaceutical research [40].

Cladribine, which targets the adenosine deaminase protein, is an active substance approved in 1993 with the integration of virtual screening and docking methods [13]. In drug repurposing studies, it has been observed that the combined use of two in silico methods results in higher accuracy compared to using each method separately [57].

5. Successes and ongoing challenges

CADD methods greatly speed up early drug discovery, allowing rational selection of drug-like compounds for therapeutic needs [58]. To ensure the success of the docking experiment, it is crucial to select the appropriate docking software and protocol [59]. Recently, platforms have been available that combine various virtual screening approaches, support parallel computing, and provide flexibility to users. By optimizing data processing, seamlessly integrating CADD software, and leveraging parallel computing architecture, we aim to enable users to complete data analytics projects faster and more efficiently [60].

According to research reports, Artificial Intelligence (AI) drug discovery firms are creating high expectations for research and development by reducing costs and accelerating timelines, once molecules are in clinical trials [61]. AI and machine learning are also used to understand the basis of diseases and evaluate the efficacy of drugs. In drug development, AI can be useful in analyzing molecular structures and identifying potential drug candidates. Predicting protein binding sites and performing structure-based virtual screens are crucial steps in the drug development process. These technologies can potentially speed up the process, reduce costs, and facilitate the discovery of more effective treatments. However, there are challenges in this area such as choosing the right dataset, model training, and result validation [62].

We can also talk about deep learning as another advancement. Deep Learning (DL) is a methodology that has been used for language and image processing for decades. However, its applications in drug discovery have emerged only in the past few years. This acceleration is attributed to the use of Graphics Processing Units [GPUs] to overcome computationally expensive calculations associated with deep learning. Compared to traditional machine learning methods, DL uses various processing layers known as neurons to make predictions based on large collections of multidimensional data. There are various types of deep learning architectures available, but prominent applications in drug design and discovery include Convolutional Neural Networks (CNNs), Recurrent Neural Networks (RNNs), Long Short-Term Memory (LSTM), and Multi-Task Learning (MTL). The creation of new drugs targeting previously unaddressed drug targets is anticipated to enhance patient care standards by increasing therapeutic efficacy and reducing side effects. With the increasing prevalence of newer techniques like deep learning in academic drug design research, a significant increase in the efficiency of conducted studies is expected [63].

Despite the significant advancements in computer-aided drug discovery, the time required to launch a new drug has not reduced significantly. Studies indicate that many active compounds found through computational techniques exhibit similar activity to compounds in clinical trials. However, most of these compounds cannot survive clinical trial stages due to the lack of necessary pharmacokinetic properties, despite demonstrating good antiviral activity and a well-defined mechanism of action. This highlights the importance of developing additional and complementary tools for evaluating pharmacokinetic properties and off-target effects. In silico approaches can provide pre-selection for candidates, while biological validation of computational predictions is also necessary. With the increasing availability of curated experimental datasets, physics-based methods and artificial intelligence techniques are expected to play a more supportive role in evaluating both the pharmacodynamic and pharmacokinetic properties of investigated compounds. Lastly, an increased probability of highly contagious disease outbreaks, underscoring the potentially significant role of successful computational strategies in combating future challenging diseases [64].

6. Conclusion

Computer-aided drug design (CADD) methods are a useful tool in the process of drug discovery. It is an area of technology that is currently developing and offers faster, more economical, and higher success rates in drug discovery. The most used CADD method is molecular docking. However, the use of multiple methods increases the success rate of the results by creating synergistic effects. The molecular docking examines the docking and interactions between the target structure and possible drug molecules. Many programs have been developed and continue to develop with technologies such as artificial intelligence, machine learning, and deep learning.

The COVID-19 pandemic has had a significant impact on the world since 2019. Although the virus is like other previously discovered members of the family, effective treatments must be developed. The rapid spread of the virus, the emergence of various variants, and the increasing number of cases and deaths have necessitated the acceleration of the drug discovery process. Multiple targets have been established in studies of the virus. Among these targets, Mpro and RdRp are two prominent proteases. These proteases are critical enzymes in virus proliferation, making it challenging to find effective drug molecules.

Drug discovery is a lengthy and exhaustive process that involves identifying new molecules that can be used to create a drug. One approach to accelerate this process is to reuse or repurpose molecules that have already been approved for other uses. This approach can help to shorten the process to some extent. Virtual screening, molecular docking, and molecular dynamics methods can be used to screen molecules in silico before further in vitro studies. This enables the interaction between molecules to be calculated on a computer.

This book chapter discusses the structural and pathological properties of the SARS-CoV-2 virus, which caused the COVID-19 pandemic. It also discusses the advantages and disadvantages of using molecular docking as a method called CADD. The chapter also covers drug discovery processes against the SARS-CoV-2 virus using molecular docking. It gives a case study and multiple examples of how molecular docking studies contribute to the prediction process and drug discovery processes. Finally, the chapter concludes with current and future expectations.

Conflict of interest

The authors declare no conflict of interest.

Appendices and nomenclature

SARS-CoV	Severe acute respiratory syndrome coronavirus
MERS-CoV	Middle East respiratory syndrome coronavirus
N	nucleocapsid protein
M	membrane protein
E	envelope protein
S	spike glycoprotein
ACE2	angiotensin-converting enzyme-2
DAMPs	damage-associated molecular patterns
RdRp	RNA-dependent RNA polymerase
Mpro	main protease
CADD	computer-aided drug design
QM	quantum mechanics
DFT	density functional theory
QSAR	quantitative structure-activity relationship
SAR	structure-activity relationship
SBDD	structure-based drug design
LBDD	ligand-based drug design
AI	artificial intelligence
DL	Deep Learning
GPUs	Graphics Processing Units
CNNs	Convolutional Neural Networks
RNNs	Recurrent Neural Networks
LSTM	Long Short-Term Memory
MTL	Multi-Task Learning

Author details

Merve Yildirim and Ismail Celik*
Faculty of Pharmacy, Department of Pharmaceutical Chemistry, Erciyes University, Kayseri, Turkey

*Address all correspondence to: ismailcelik@erciyes.edu.tr

References

[1] Hu B, Guo H, Zhou P, Shi ZL. Characteristics of SARS-CoV-2 and COVID-19. Nature Reviews. Microbiology. 2021;**19**(3):141-154

[2] Gordon DE, Jang GM, Bouhaddou M, Xu J, Obernier K, White KM, et al. A SARS-CoV-2 protein interaction map reveals targets for drug repurposing. Nature. 2020;**583**(7816):459-468

[3] Estrada E. Topological analysis of SARS CoV-2 main protease. Chaos. 2020;**30**(6):061102

[4] Tang D, Comish P, Kang R. The hallmarks of COVID-19 disease. PLOS Pathogens. 2020;**16**(5):e1008536

[5] Li G, Fan Y, Lai Y, Han T, Li Z, Zhou P, et al. Coronavirus infections and immune responses. Journal of Medical Virology. 2020;**92**(4):424-432

[6] Khade SM, Yabaji SM, Srivastava J. An update on COVID-19: SARS-CoV-2 life cycle, immunopathology, and BCG vaccination. Preparative Biochemistry & Biotechnology. 2021;**51**(7):650-658

[7] Tseng YY. Modeling the effects of SARS-CoV-2 infection on the mTOR signaling pathway. In: 2022 IEEE 22nd International Conference on Bioinformatics and Bioengineering (BIBE). Taichung, Taiwan; 7-9 Nov. 2022

[8] Andola P, Pagag J, Laxman D, Guruprasad L. Fragment-based inhibitor design for SARS-CoV2 main protease. Structural Chemistry. 2022;**33**(5):1467-1487

[9] Kumar V. Understanding the complexities of SARS-CoV2 infection and its immunology: A road to immune-based therapeutics. International Immunopharmacology. 2020;**88**:106980

[10] Satapathy P, Kumar P, Gupta JK, Rabaan AA, Al Kaabi NA, Mohanty D, et al. The emergence and implications of SARS-CoV-2 omicron subvariant BA.2.86 on global health. International Journal of Surgery. 11 Jan 2024. DOI: 10.1097/JS9.0000000000001070

[11] Grover V, Kumar V, Puri V, Jain A. Chlorhexidine and SARS-CoV-2 main protease: Molecular docking study. J Indian Soc Periodontol. 2022;**26**(5):512-514

[12] Mohamed EAR, Abdel-Rahman IM, Zaki MEA, Al-Khdhairawi A, Abdelhamid MM, Alqaisi AM, et al. In silico prediction of potential inhibitors for SARS-CoV-2 omicron variant using molecular docking and dynamics simulation-based drug repurposing. Journal of Molecular Modeling. 2023;**29**(3):70

[13] Sabe VT, Ntombela T, Jhamba LA, Maguire GEM, Govender T, Naicker T, et al. Current trends in computer aided drug design and a highlight of drugs discovered via computational techniques: A review. European Journal of Medicinal Chemistry. 2021;**224**:113705

[14] Khalifa HO, Al Ramahi YM. After the hurricane: Anti-COVID-19 drugs development, molecular mechanisms of action and future perspectives. International Journal of Molecular Sciences. 2024;**25**(2):739

[15] Tang Y, Zhu W, Chen K, Jiang H. New technologies in computer-aided drug design: Toward target identification and new chemical entity discovery.

Drug Discovery Today: Technologies. 2006;**3**(3):307-313

[16] Hosseini M, Chen W, Xiao D, Wang C. Computational molecular docking and virtual screening revealed promising SARS-CoV-2 drugs. Precision Clinical Medicine. 2021;**4**(1):1-16

[17] Hashem HE, Ahmad S, Kumer A, Bakri YE. In silico and in vitro prediction of new synthesized N-heterocyclic compounds as anti-SARS-CoV-2. Scientific Reports. 2024;**14**(1):1152

[18] Lohning AE, Levonis SM, Williams-Noonan BJ, Schweiker SS. A practical guide to molecular docking and homology modelling for medicinal chemists. Current Topics in Medicinal Chemistry. 2017;**17**(18):2023-2040

[19] Gajjar ND, Dhameliya TM, Shah GB. In search of RdRp and Mpro inhibitors against SARS CoV-2: Molecular docking, molecular dynamic simulations and ADMET analysis. Journal of Molecular Structure. 2021;**1239**:130488

[20] De Sousa-Coelho AL, Fraqueza G, Aureliano M. Repurposing therapeutic drugs complexed to vanadium in cancer. Pharmaceuticals. 2024;**17**(1):12

[21] Anderson C, Bucholc M, McClean PL, Zhang S-D. The potential of a stratified approach to drug repurposing in Alzheimer’s disease. Biomolecules. 2024;**14**(1):11

[22] Bogacheva MS, Kuivanen S, Potdar S, Hassinen A, Huuskonen S, Pöhner I, et al. Drug repurposing platform for deciphering the druggable SARS-CoV-2 interactome. Antiviral Research. 2024;**223**:105813

[23] Zhang XY, Huang HJ, Zhuang DL, Nasser MI, Yang MH, Zhu P, et al. Biological, clinical and epidemiological features of COVID-19, SARS and MERS and AutoDock simulation of ACE2. Infectious Diseases of Poverty. 2020;**9**(1):99

[24] Trott O, Olson AJ. AutoDock Vina: Improving the speed and accuracy of docking with a new scoring function, efficient optimization, and multithreading. Journal of Computational Chemistry. 2010;**31**(2):455-461

[25] Rarey M, Kramer B, Lengauer T, Klebe G. A fast flexible docking method using an incremental construction algorithm. Journal of Molecular Biology. 1996;**261**(3):470-489

[26] Vilar S, Cozza G, Moro S. Medicinal chemistry and the molecular operating environment (MOE): Application of QSAR and molecular docking to drug discovery. Current Topics in Medicinal Chemistry. 2008;**8**(18):1555-1572

[27] Rani I, Kalsi A, Kaur G, Sharma P, Gupta S, Gautam RK, et al. Modern drug discovery applications for the identification of novel candidates for COVID-19 infections. Annals of Medicine and Surgery (London). 2022;**80**:104125

[28] Friesner RA, Murphy RB, Repasky MP, Frye LL, Greenwood JR, Halgren TA, et al. Extra precision glide: Docking and scoring incorporating a model of hydrophobic enclosure for protein−ligand complexes. Journal of Medicinal Chemistry. 2006;**49**(21):6177-6196

[29] Lokhande KB, Doiphode S, Vyas R, Swamy KV. Molecular docking and simulation studies on SARS-CoV-2 M(pro) reveals Mitoxantrone, Leucovorin, Birinapant, and Dynasore as potent drugs against COVID-19. Journal of Biomolecular Structure & Dynamics. 2021;**39**(18):7294-7305

[30] Ouassaf M, Belaidi S, Chtita S, Lanez T, Abul Qais F, Md AH. Combined molecular docking and dynamics simulations studies of natural compounds as potent inhibitors against SARS-CoV-2 main protease. Journal of Biomolecular Structure and Dynamics. 2022;**40**(21):11264-11273

[31] Ahamad S, Gupta D, Kumar V. Targeting SARS-CoV-2 nucleocapsid oligomerization: Insights from molecular docking and molecular dynamics simulations. Journal of Biomolecular Structure & Dynamics. 2022;**40**(6):2430-2443

[32] Alazmi M, Motwalli O. In silico virtual screening, characterization, docking and molecular dynamics studies of crucial SARS-CoV-2 proteins. Journal of Biomolecular Structure & Dynamics. 2021;**39**(17):6761-6771

[33] Khelfaoui H, Harkati D, Saleh BA. Molecular docking, molecular dynamics simulations and reactivity, studies on approved drugs library targeting ACE2 and SARS-CoV-2 binding with ACE2. Journal of Biomolecular Structure & Dynamics. 2021;**39**(18):7246-7262

[34] Zamzami MA. Molecular docking, molecular dynamics simulation and MM-GBSA studies of the activity of glycyrrhizin relevant substructures on SARS-CoV-2 RNA-dependent-RNA polymerase. Journal of Biomolecular Structure and Dynamics. 2023;**41**(5):1846-1858

[35] Stanzione F, Giangreco I, Cole JC. Use of molecular docking computational tools in drug discovery. Progress in Medicinal Chemistry. 2021;**60**:273-343

[36] Li J, Fu A, Zhang L. An overview of scoring functions used for protein-ligand interactions in molecular docking. Interdisciplinary Sciences. 2019;**11**(2):320-328

[37] Fan M, Wang J, Jiang H, Feng Y, Mahdavi M, Madduri K, et al. GPU-accelerated flexible molecular docking. The Journal of Physical Chemistry B. 2021;**125**(4):1049-1060

[38] Hessel SS, Dwivany FM, Zainuddin IM, Wikantika K, Celik I, Emran TB, et al. A computational simulation appraisal of banana lectin as a potential anti-SARS-CoV-2 candidate by targeting the receptor-binding domain. Journal, Genetic Engineering & Biotechnology. 2023;**21**(1):148

[39] Rogers DM, Agarwal R, Vermaas JV, Smith MD, Rajeshwar RT, Cooper C, et al. SARS-CoV2 billion-compound docking. Scientific Data. 2023;**10**(1):173

[40] Meng X-Y, Zhang H-X, Mezei M, Cui M. Molecular docking: A powerful approach for structure-based drug discovery. Current Computer-Aided Drug Design. 2011;**7**(2):146-157

[41] Agarwal N, Fatima A, Bhattacharya P, Muthu S, Arora H, Siddiqui N, et al. Evaluation of experimental, computational, molecular docking and dynamic simulation of flucytosine. Journal of Biomolecular Structure & Dynamics. 2023;**41**(20):10430-10449

[42] Lebedeva NS, Gubarev YA, Mamardashvili GM, Zaitceva SV, Zdanovich SA, Malyasova AS, et al. Theoretical and experimental study of interaction of macroheterocyclic compounds with ORF3a of SARS-CoV-2. Scientific Reports. 2021;**11**(1):19481

[43] Shree P, Mishra P, Selvaraj C, Singh SK, Chaube R, Garg N, et al. Targeting COVID-19 (SARS-CoV-2) main protease through active

phytochemicals of ayurvedic medicinal plants - Withania somnifera (Ashwagandha), Tinospora cordifolia (Giloy) and Ocimum sanctum (Tulsi) - a molecular docking study. Journal of Biomolecular Structure & Dynamics. 2022;**40**(1):190-203

[44] Alagu Lakshmi S, Shafreen RMB, Priya A, Shunmugiah KP. Ethnomedicines of Indian origin for combating COVID-19 infection by hampering the viral replication: Using structure-based drug discovery approach. Journal of Biomolecular Structure & Dynamics. 2021;**39**(13):4594-4609

[45] Ashraf-Uz-Zaman M, Chua TK, Li X, Yao Y, Moku BK, Mishra CB, et al. Design, synthesis, X-ray crystallography, and biological activities of covalent, non-Peptidic inhibitors of SARS-CoV-2 Main protease. ACS Infectious Diseases. 2024;**10**(2):715-731

[46] Agost-Beltrán L, de la Hoz-Rodríguez S, Bou-Iserte L, Rodríguez S, Fernández-de-la-Pradilla A, González FV. Advances in the development of SARS-CoV-2 Mpro inhibitors. Molecules. 2022;**27**(8):2523

[47] Malone BF, Perry JK, Olinares PDB, Lee HW, Chen J, Appleby TC, et al. Structural basis for substrate selection by the SARS-CoV-2 replicase. Nature. 2023;**614**(7949):781-787

[48] Noske GD, de Souza SE, de Godoy MO, Dolci I, Fernandes RS, Guido RVC, et al. Structural basis of nirmatrelvir and ensitrelvir activity against naturally occurring polymorphisms of the SARS-CoV-2 main protease. Journal of Biological Chemistry. 2023;**299**(3):103004

[49] Mao L, Shaabani N, Zhang X, Jin C, Xu W, Argent C, et al. Olgotrelvir, a dual inhibitor of SARS-CoV-2 Mpro and cathepsin L, as a standalone antiviral oral intervention candidate for COVID-19. Medicine. 2024;**5**(1):42-61.e23

[50] Zhang L, Lin D, Sun X, Curth U, Drosten C, Sauerhering L, et al. Crystal structure of SARS-CoV-2 main protease provides a basis for design of improved α-ketoamide inhibitors. Science. 2020;**368**(6489):409-412

[51] Abas A, Tallei T, Fatimawali F, Celik I, Alhumaydhi F, Emran T, et al. 4?-fluorouridine as a potential COVID-19 oral drug?: A review [version 2; peer review: 1 approved with reservations, 1 not approved]. F1000Research. 2023;**11**(410)

[52] Joyce RP, Hu VW, Wang J. The history, mechanism, and perspectives of nirmatrelvir (PF-07321332): An orally bioavailable main protease inhibitor used in combination with ritonavir to reduce COVID-19-related hospitalizations. Medicinal Chemistry Research. 2022;**31**(10):1637-1646

[53] Owen DR, Allerton CMN, Anderson AS, Aschenbrenner L, Avery M, Berritt S, et al. An oral SARS-CoV-2 M^{pro} inhibitor clinical candidate for the treatment of COVID-19. Science. 2021;**374**(6575):1586-1593

[54] Chtita S, Belaidi S, Qais FA, Ouassaf M, AlMogren MM, Al-Zahrani AA, et al. Unsymmetrical aromatic disulfides as SARS-CoV-2 Mpro inhibitors: Molecular docking, molecular dynamics, and ADME scoring investigations. Journal of King Saud University - Science. 2022;**34**(7):102226

[55] Yang X, Xing X, Liu Y, Zheng Y. Screening of potential inhibitors targeting the main protease structure of SARS-CoV-2 via molecular docking. Frontiers in Pharmacology. 2022;**13**

[56] Prada-Gracia D, Huerta-Yépez S, Moreno-Vargas LM. Application of computational methods for anticancer drug discovery, design, and optimization. Boletín Médico del Hospital Infantil de México. 2016;**73**(6):411-423

[57] Yuda GPWC, Hanif N, Hermawan A. Computational screening using a combination of ligand-based machine learning and molecular docking methods for the repurposing of antivirals targeting the SARS-CoV-2 Main protease. DARU. Journal of Pharmaceutical Sciences. 2023:2008-2231

[58] Bassani D, Moro S. Past, present, and future perspectives on computer-aided drug design methodologies. Molecules. 2023;**28**(9):3906

[59] Paul SK, Guendouzi A, Banerjee A, Guendouzi A, Haldar R. Identification of approved drugs with ALDH1A1 inhibitory potential aimed at enhancing chemotherapy sensitivity in cancer cells: An in-silico drug repurposing approach. Journal of Biomolecular Structure and Dynamics. 2024;**0**(0):1-15

[60] Khan MF, Kandwal S, Fayne D. DataPype: A fully automated unified software platform for computer-aided drug design. ACS Omega. 2023;**8**(42):39468-39480

[61] Jayatunga MKP, Xie W, Ruder L, Schulze U, Meier C. AI in small-molecule drug discovery: A coming wave? Nature Reviews. Drug Discovery. 2022;**21**(3):175-176

[62] Seo S, Lee JW. Applications of big data and AI-driven technologies in CADD (computer-aided drug design). In: Gore M, Jagtap UB, editors. Computational Drug Discovery and Design. New York, NY: Springer US; 2024. pp. 295-305

[63] Dorahy G, Chen JZ, Balle T. Computer-aided drug design towards new psychotropic and neurological drugs. Molecules. 2023;**28**(3):1324

[64] Pavan M, Moro S. Lessons learnt from COVID-19: Computational strategies for facing present and future pandemics. International Journal of Molecular Sciences. 2023;**24**(5):4401

Chapter 6

Personalised Medicine and Molecular Docking: Tailoring Drug Discovery for Individual Patients

Noopur Khare and Pragati Khare

Abstract

The combination of molecular docking with personalised medicine represents a paradigm shift in drug development, providing unmatched accuracy in customising therapeutic approaches for specific patients. This collaborative effort utilises cutting-edge computational methods, including molecular docking, in conjunction with genetic insights to optimise and anticipate drug-receptor interactions. Revolutionary achievements could be further amplified by integrating large-scale omics data, artificial intelligence, and structural biology discoveries. Molecular docking and personalised medicine are developing fields that could lead to treatments that take into account each patient's unique molecular profile in addition to previously unheard-of levels of accuracy in disease diagnosis. This revolutionary landscape will be further enhanced by future developments in quantum computing, CRISPR-based gene editing, and biomarker discovery. These advances will enable the realisation of a healthcare paradigm in which interventions are not only precise but also proactive, thereby realising the full potential of customised therapeutic strategies for improved patient outcomes.

Keywords: personalised medicine, molecular docking, drug discovery, individualised therapeutics, genomic insights

1. Introduction

A paradigm shift is occurring in the field of contemporary healthcare, bringing in an era where patients' individual genetic composition will be precisely taken into account when designing medical therapies. The personalised medicine method is a new strategy that acknowledges the inherent variety in individuals' genetic codes and biological reactions [1]. It deviates from the standard one-size-fits-all model. The symbiotic relationship between molecular docking, a complex dance of genetics and computational science, and personalised medicine, as we stand at the intersection of biology, technology, and medicine. This relationship has the potential to turn drug discovery into a highly individualised endeavour [2]. Understanding the genetic underpinnings of health and disease is the cornerstone of personalised medicine. The complexity of the human genome has been revealed by advances in genomics, which have also revealed a tapestry of genetic variations that underpin disease vulnerability and impact responses to therapeutic interventions. Simultaneously, the advent of omics technologies, which include transcriptomics, proteomics, metabolomics, and genomes, has enabled researchers to interpret the molecular subtleties that differentiate individual patients. Personalised medicine makes use of this abundance of data to divide patients into groups according to their distinct molecular profiles, which enables the development of more focused and efficient treatment plans [3].

In addition to this novel strategy, molecular docking—a computational method with roots in structural biology—has become a key component in the drug development process. Drugs with improved specificity and efficacy can be logically designed by molecular docking, which simulates the interaction between tiny drug-like molecules and their biological targets. The intersection of molecular docking and personalised medicine, examining how the combination of patient-specific genetic data and computational approaches is changing the face of drug discovery [4]. We take a close look at issues and possibilities for the future as well as an insightful look at case studies to explore a future in which medication is not only customised for each patient but also carefully built for their unique genetic makeup.

The complex interactions between molecular docking and personalised medicine, revealing how these two fields can work together to transform healthcare. The story will develop as we explore the methods and uses of molecular docking to demonstrate how this computational power may be used to create medications that perfectly fit each patient's unique molecular makeup. Drug discovery procedures that use patient-specific genetic data have the potential to revolutionise healthcare by promoting preventative and predictive medicine in addition to improving treatment efficacy. This chapter attempts to add to the ongoing discussion on moving personalised medicine and molecular docking into a new frontier in healthcare innovation by exploring ethical considerations, technical hurdles, and potential trajectories [5].

2. Foundations of personalised medicine

2.1 Genetic basis of personalised medicine

2.1.1 Role of genomics in understanding individual variations

A thorough grasp of the genetic mosaic that sets each person apart is fundamental to personalised therapy. The study of an organism's complete gene pool, or genomics, offers a thorough road map for figuring out the complexities of individual genetic variants and how they affect health and disease. The human genome, which is made up of between 20,000–25,000 genes, acts as a database that determines a person's biological composition. In-depth discussion of genomics' function as the foundation of personalised medicine, along with an explanation of how genetic data is used to identify individual differences and adjust medical interventions accordingly. Finding genetic markers linked to treatment responses and illness risk is a crucial task for genomics. Thanks to developments in high-throughput sequencing technologies, it is now possible to efficiently and economically decode a person's whole genome, which enables scientists to identify gene variants that may predispose people to particular diseases. The abundance of genetic data makes it easier to divide patients into groups according to genetic characteristics they have in common, opening the door to more specialised and efficient treatment plans [6].

The investigation of individual genetic variants covers drug metabolism and response in addition to illness propensity. A branch of genomics called pharmacogenomics studies how a person's genetic makeup affects their response to medication, including how effective it is and how likely it is to cause side effects [7]. Clinicians can prescribe drugs that are not just suited to a patient's particular ailment but also best suited to their individual genetic composition by having a thorough understanding of these genetic subtleties. The genetic strands that weave the story of individual differences as we traverse the complex terrain of genomics in personalised medicine. We set out on a trip that emphasises the revolutionary potential of deciphering the genetic code for customising medical therapies to the unique needs of each patient through a thorough assessment of the role of genomics in healthcare (**Figure 1**) [8].

2.1.2 Identification of genetic markers for disease susceptibility

Knowing the genetic foundations of illness susceptibility is essential to the goal of personalised medicine. Numerous genetic markers—distinct variations in DNA sequences—have been found through genomic research. These markers serve as markers that help researchers better understand an individual's susceptibility to particular diseases [9]. This part delves into the complex terrain of determining genetic markers associated with illness vulnerability, highlighting the revolutionary implications of this understanding on customising medicinal therapies. Single nucleotide polymorphisms (SNPs) and more complicated structural variants are examples of genetic markers that act as genomic fingerprints to identify individuals. By utilising sophisticated sequencing methods and conducting extensive genome-wide association studies (GWAS), scientists are able to identify the markers linked to heightened vulnerability to a range of illnesses. These genetic discoveries have significant ramifications for early detection, risk assessment, and preventive measures [10].

In multifactorial disorders, when the total risk is influenced by both genetic and environmental variables, the identification of genetic markers is very clear. Complex

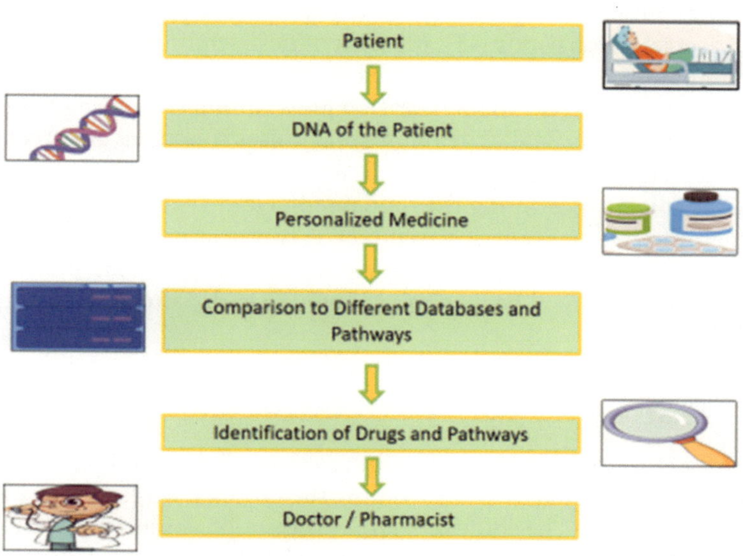

Figure 1.
Genetic basis of personalised medicine.

genetic architectures are present in diseases like cancer, cardiovascular disease, and neurological illnesses. By identifying particular markers, patients can be categorised according to their genetic predispositions. In addition to improving prognosis, this stratification makes it easier to design tailored therapies that take unique risk profiles into account [11]. Personalised medicine is becoming more and more dependent on the ability to detect genetic markers for disease vulnerability. Clinicians can now proactively tailor prevention strategies, surveillance protocols, and therapeutic approaches by deciphering the genetic clues embedded in an individual's DNA. This marks a new era in healthcare where interventions are precisely aligned with each patient's unique genetic makeup [12].

2.2 Omics technologies

2.2.1 Genomics

With genomics at its core, omics technologies are revolutionising the field of personalised medicine by enabling the deciphering of the intricate workings of individual genetic codes. Genomics is a branch of omics that studies all of an organism's genes in detail. It offers a thorough understanding of the genetic basis of health, disease susceptibility, and response to therapy [13]. Utilising high-throughput sequencing technology, like next-generation sequencing (NGS), which makes it possible to quickly and affordably sequence complete genomes, is at the core of genomics research. With the advancement of technology, scientists can now quickly decode the three billion base pairs that make up the human genome, bringing genomics into the mainstream of scientific study [14].

The identification of genetic differences, such as single nucleotide polymorphisms (SNPs) and structural changes, among heterogeneous populations is largely dependent on genomics. Large-scale programmes such as the Human Genome Project and others have accumulated a vast store of genomic data that makes it easier to identify genetic markers linked to treatment results and disease risk. In the effort to decipher the complex connection between genetics and personal health, these markers act as navigational aids [15]. Moreover, the discipline of pharmacogenomics, which studies how genetic variants affect drug responses, benefits from the extension of genomics. By customising medication prescriptions according to a person's genetic composition, this field of genomics maximises treatment effectiveness and reduces side effects [16]. Genomic research serves as a cornerstone in the era of personalised medicine by offering the fundamental knowledge needed to understand the subtleties of individual genetic variants. The future of precision medicine and better patient outcomes is promised by the integration of genomics into healthcare, where medical therapies will be precisely tailored to each patient's distinct genetic signature [17].

2.2.2 Transcriptomics

Transcriptomics is a dynamic field within the complicated field of omics technologies that is essential to understanding the intricate workings of gene expression within an organism. The study of all RNA transcripts—molecules that transfer genetic information from DNA to protein synthesis—is the focus of transcriptomics [18]. This field provides insights into the molecular orchestration of biological processes by helping to understand how genes are activated or silenced under different physiological situations. Transcriptomics has undergone a revolution thanks to high-throughput methods like RNA sequencing (RNA-Seq), which allow for the simultaneous investigation of thousands of transcripts from a variety of tissues and cell types. With the help of this technique, the dynamic transcriptome may be fully captured, illuminating the subtleties of non-coding RNA expression, alternative splicing, and gene regulation [19].

Transcriptomics reveals the molecular markers linked to certain diseases and the unique responses to treatment interventions, which has profound implications for personalised medicine. Transcriptomics helps to stratify patient groups and directs doctors in the selection of customised treatment methods by detecting gene expression patterns indicative of disease states or treatment efficacy. Transcriptomics is essentially a molecular storyteller that tells the complex stories of gene activity in cells. Transcriptomics, a key element of omics technologies, advances our knowledge of the intricate molecular processes underlying health and disease and opens the door to more specialised and individualised medical care [20].

2.2.3 Proteomics

Proteomics is the dominant field in the mosaic of omics technologies, exploring the in-depth analysis of an organism's whole protein repertoire. Proteomics is essential for understanding the dynamic landscape of protein expression, changes, and interactions within cells. Proteins are the molecular workhorses that carry out the functions encoded by the genome. Proteomic investigations are made possible by sophisticated methods like mass spectrometry and two-dimensional gel electrophoresis, which allow thousands of proteins to be identified and quantified at once. A comprehensive picture of the proteome is offered by this high-throughput method, which captures the nuances of cellular functions, signalling cascades, and reactions to external stimuli [21]. By revealing protein patterns linked to certain diseases, proteomics makes a substantial contribution to personalised medicine by facilitating the discovery of biomarkers for early diagnosis and prognosis. Additionally, it clarifies the range of protein isoforms and post-translational changes, providing information on how therapy responses differ throughout individuals [22]. Proteomics' incorporation into biomedical research enables a more thorough comprehension of the complex molecular mechanisms underlying health and illness. Proteomics, a fundamental component of omics technologies, opens the door for precision medicine, which promises more efficient and individualised treatment plans by customising therapeutic approaches to each patient's own protein composition [23].

2.2.4 Metabolomics

Within the complex field of omics technologies, metabolomics is a potent field devoted to the in-depth analysis of an organism's metabolome, or the entire collection of tiny molecules, or metabolites, that are involved in cellular functions. A snapshot of the physiological state is given in real time by metabolomics, which captures the results of cellular activity and their dynamic interactions [24]. Metabolomic analyses are powered by state-of-the-art analytical methods including mass spectrometry and nuclear magnetic resonance (NMR) spectroscopy, which allow for the identification and measurement of a wide range of metabolites. This high-throughput method provides information on biochemical processes, cellular homeostasis, and metabolic pathways [25].

Metabolomics plays a major role in personalised medicine by providing insights into the metabolic fingerprints linked to different diseases and how each patient reacts to different therapies. Metabolomics profiles the distinct fingerprint of metabolites in biological samples to help identify biomarkers, diagnose diseases, and optimise treatment approaches. Metabolomics, a crucial component of omics technologies, completes the whole picture of an organism's molecular composition [26]. Enhancing our knowledge of biochemical networks and advancing precision medicine—where treatments are customised to each patient's unique metabolic profile—are two benefits of this research. The addition of metabolomics to the omics toolkit represents a paradigm change in healthcare towards more individualised and complex interventions (**Figure 2**) [27].

2.3 Patient stratification and targeted therapies

2.3.1 Subpopulations and treatment responses

The deliberate identification and grouping of patients into subpopulations according to shared genetic, molecular, or clinical traits is a fundamental component of personalised medicine. A more sophisticated knowledge of individual differences in disease susceptibility and treatment responses is made possible by this patient stratification. Targeted medicines can be carefully adjusted to satisfy the unique needs of different subgroups by taking into account the heterogeneity within patient cohorts [28]. Patient stratification in the era of genetic medicine greatly depends on the knowledge obtained from omics technology. Subpopulations with differing susceptibilities to diseases and treatment responses are defined by molecular signatures that are identified through the use of genomic profiling, transcriptomics, proteomics,

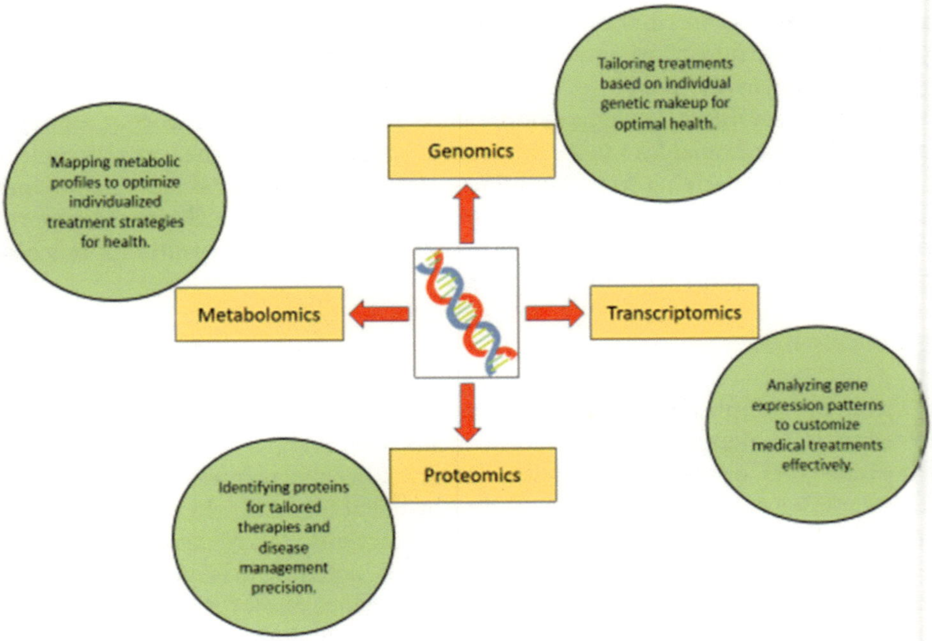

Figure 2.
Foundations of personalised medicine & omics technologies.

and metabolomics. This method goes beyond the conventional one-size-fits-all paradigm by acknowledging the distinct genetic composition that impacts each person's particular health trajectory [29].

When it comes to targeted medicines, patient classification is very important. These treatments target certain pathways or substances linked to disease processes, acting at the molecular or cellular level. Targeted therapies, which represent a paradigm shift from conventional treatments, can achieve higher efficacy and fewer adverse effects by identifying subpopulations with shared molecular markers. Patient stratification is a significant step towards precision medicine, where treatment approaches are tailored to each patient's unique needs. This trend is being driven by advancements in omics technologies. This customised approach not only improves treatment results but also establishes the framework for a more efficient and individualised healthcare system [30].

2.3.2 Tailoring therapies based on patient characteristics

The emergence of personalised medicine signals a paradigm shift in healthcare by highlighting the necessity of customising treatment procedures according to the unique characteristics of each patient. One of the main tenets of this paradigm is patient stratification, which is the methodical division of patients into groups according to their distinct clinical, molecular, or genetic characteristics [31]. In addition to revealing the innate variation among patient populations, this stratification paves the way for the creation of tailored targeted medicines that target particular traits unique to each grouping. Customising treatment plans according to the features of each patient represents a significant shift from the traditional one-size-fits-all approach. By merging data from omics technologies—genomics, transcriptomics, and proteomics, for example—clinicians can better comprehend the intricate molecular mechanisms underlying disease and how treatments affect patients' outcomes. This information facilitates the discovery of biomarkers, genetic variants, and other characteristics unique to each patient that influence the choice of targeted treatments [32].

Personalised medicine is typified by targeted medicines, which precisely target specific molecules or pathways involved in disease processes at the molecular or cellular level. These treatments represent a paradigm change towards more effective and individualised healthcare by matching treatment plans to the unique features of patient subgroups [33]. They also increase efficacy while reducing side effects. The combination of targeted therapies and patient stratification is a novel approach that allows medical interventions to be customised to each patient's specific characteristics. This is ushering in a new era of precision medicine that has the potential to completely transform patient care and treatment outcomes [34].

3. Molecular docking in drug discovery

3.1 Overview of molecular docking

A key component of rational drug design is molecular docking, which offers a computer framework for predicting the favoured orientation and binding affinity of a small molecule (ligand) inside the target macromolecule's (receptor) active site. This method is essential for optimising the interactions between drugs and receptors, which makes it easier to design and create new pharmaceutical agents [35]. Fundamentally, molecular docking is based on scoring functions and algorithms that assess the geometry and energetics of putative ligand-receptor complexes. The procedure entails methodically examining the conformational space of the receptor and ligand in order to forecast the ideal binding pose in light of complimentary structural characteristics. In order to model and evaluate the stability of the ligand-receptor complex, the concepts of molecular docking take into account intermolecular forces such as hydrogen bonding, electrostatic interactions, and van der Waals forces [36].

Clarifying the binding mechanism and affinity between a drug candidate and its target is the main goal of molecular docking, since it provides important information about the viability and possible effectiveness of the interaction. By reducing the number of possible candidates for additional experimental validation, this *in silico* method greatly speeds up the drug discovery process and ultimately leads to the production of therapeutic medicines with improved specificity and affinity for their target targets (**Figure 3**) [37].

3.1.1 Importance in rational drug design

Molecular docking is essential to the logical design of pharmaceuticals because it acts as a computerised compass to navigate the complex process from lead identification to the creation of potent medicinal compounds. Molecular docking, at its core, offers a systematic and economical way to anticipate and optimise the interactions between small compounds and their target macromolecules, which is essential to rational drug design. Because it can virtually screen large compound libraries, molecular docking is important because it can speed up the drug discovery process [38]. Docking algorithms anticipate the most favourable binding conformations and energetically favourable interactions by modelling the binding of putative drug candidates to target receptors. Promising lead compounds are prioritised, their structures are optimised, and experimental efforts are rationalised for additional validation with the use of this prediction capability [39].

Additionally, the investigation of structure-activity relationships (SAR) and the comprehension of ligand-receptor interactions at the atomic level are made easier by molecular docking. It guides the iterative process of drug optimisation by enabling researchers to fine-tune chemical structures to improve binding affinities and specificity [40]. Molecular docking, which serves as a computational link between

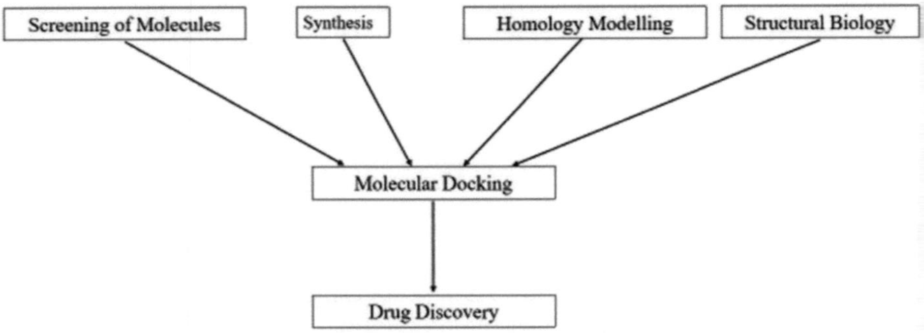

Figure 3.
Molecular docking.

theoretical predictions and experimental validations, is fundamental to rational drug design. In addition to hastening the identification of possible treatments, its function in optimising the drug discovery process helps create safer, more effective medications that have a better chance of succeeding in clinical settings [41].

3.2 Techniques and algorithms

3.2.1 Ligand-based docking

A basic method of molecular docking that emphasises the features and attributes of the small molecule ligand is called "ligand-based docking." This method is predicated on the idea that ligands with similar structures will have comparable binding patterns and bioactivities. In contrast to structure-based docking, which depends on the structure of the target receptor, ligand-based docking is useful in situations where experimental structures are not available and does not require receptor structural information [42]. In the process of ligand-based docking, algorithms compare the chemical structures of known ligands and the possible drug candidate in order to estimate the binding affinity based on properties that are common [43]. Molecular fingerprinting, pharmacophore modelling, and quantitative structure-activity relationship (QSAR) analysis are examples of common approaches. Scoring functions are used by molecular docking programmes like AutoDock and GOLD to rate possible ligand conformations and orientations within the binding site [44].

Particularly useful is ligand-based docking when the receptor structure is unclear or difficult to ascertain. It guides the search for possible therapeutic candidates by utilising the body of knowledge already available regarding ligand-receptor interactions and chemical similarities. Before moving on to more resource-intensive experimental validation, this method helps identify and optimise lead compounds with desirable pharmacological properties, which is crucial in the early phases of drug discovery [45].

3.2.2 Structure-based docking

A pillar of the molecular docking community is structure-based docking, which depends on the three-dimensional structures of the target receptor and the tiny ligand. Using computer algorithms, this method predicts the ideal binding shape and affinity between the ligand and the receptor, offering vital information for well-thought-out drug design. In structure-based docking, the search for ligand binding sites is directed by the crystallographic or computationally predicted structure of the target receptor [46]. The spatial orientations of the ligand within the binding pocket of the receptor are carefully explored using algorithms like the Lamarckian Genetic Algorithm and Monte Carlo simulations. To help determine the best binding poses, the scoring functions evaluate the energetics and complementarity of possible ligand-receptor complexes. This method works best in situations where comprehensive structural data about the target is accessible, allowing for an accurate comprehension of ligand-receptor interactions at the atomic level. It is essential for virtual screening of big compound libraries and lead optimisation. Structure-based docking is a valuable tool in drug development since it facilitates the identification of promising candidates for treatment and provides valuable insights for further experiments, ultimately leading to the rational design of pharmacological medicines with improved efficacy and specificity (**Figure 4**) [47].

3.3 Applications in drug discovery

3.3.1 Identification of potential drug candidates

Drug discovery benefits greatly from the use of molecular docking, especially in the early phases when finding viable treatment options is crucial. By predicting the binding affinity and orientation of tiny molecules within the active region of a target receptor, docking algorithms enable the systematic screening of large chemical libraries. Molecular docking helps find drugs that have good interactions with the target by means of virtual screening. This procedure speeds up the drug development process a great deal and allows researchers to concentrate their efforts on a small number of molecules that have the best chance of becoming successful. The algorithms enable the identification of lead candidates for additional experimental validation by prioritising compounds according to their estimated binding affinities [48].

Additionally, molecular docking guides the design and optimisation of new compounds with improved binding capabilities, assisting in the exploration of chemical space. It helps with the repeated improvement of chemical structures to increase specificity and efficacy by offering insightful information on the structure-activity relationships (SAR) of ligands. The use of molecular docking to find viable drug candidates is consistent with the rational drug design principles; it facilitates lead discovery and aids in the creation of novel pharmaceuticals that may prove to be therapeutically effective [49].

3.3.2 Optimization of drug-receptor interactions

Drug development relies heavily on molecular docking, which makes it easier to optimise how medications interact with their target receptors. Molecular docking algorithms are used to anticipate and fine-tune the binding modes, orientations, and affinities inside the target's active site after possible drug candidates have been found. To improve the effectiveness and specificity of pharmaceutical agents, rational drug

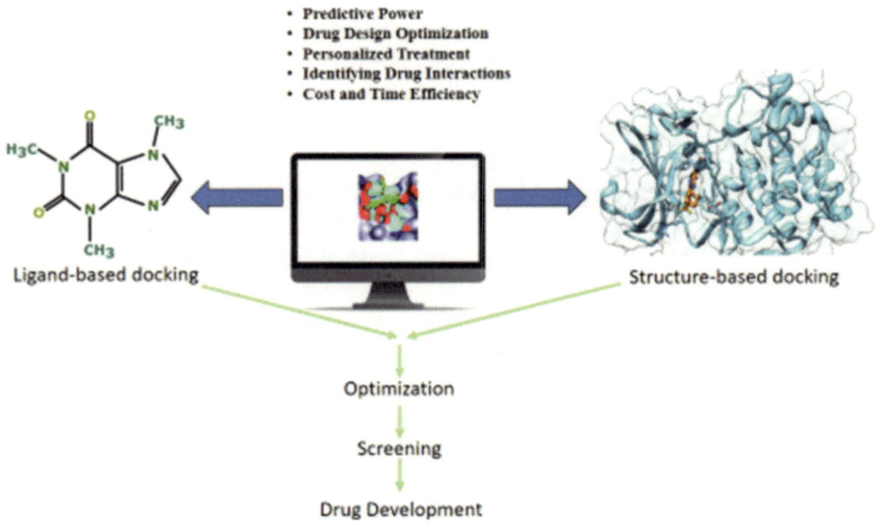

Figure 4.
Molecular docking in personalised medicine.

design includes optimising drug-receptor interactions as a critical step. Docking algorithms estimate the best binding poses by repeatedly exploring the ligand's conformational space. Medicinal chemists use these predictions as a guidance when fine-tuning drug candidates' chemical structures in order to increase binding affinity and maximise interactions with the target [50].

Moreover, molecular docking helps to clarify structure-activity relationships (SAR) by illuminating how modifications to a molecule's structure affect binding. With this information, researchers can more effectively alter drug candidates to achieve better pharmacokinetics and less off-target effects. Molecular docking expedites the drug optimisation process and lowers lead development time and costs by fusing computational predictions with experimental confirmation. In the pursuit of novel and focused pharmaceutical treatments, this application makes sure that possible medicines are not only found but also optimised for their binding interactions. This ultimately leads to the development of safer, more effective medications [51].

4. Integration of personalised medicine and molecular docking

4.1 Customised drug design

4.1.1 Incorporating patient-specific genetic information

The combination of molecular docking and personalised medicine has the potential to transform drug design through the use of genetic information specific to each patient. This methodology recognises the innate genetic variability among individuals and utilises omics data, encompassing genomics, to customise medication interventions with unparalleled accuracy. In this setting, molecular docking acts as a computational bridge, making predictions about how individual differences may affect drug-receptor interactions based on patient-specific genetic data. Docking algorithms can mimic the binding affinity and orientation of medication candidates within the setting of particular genetic markers or variants by taking into account the individual genetic composition of each patient. This customised strategy goes beyond traditional paradigms for medication design. It makes it possible to develop pharmaceuticals that are tailored to each patient's unique genetic profile, potentially increasing therapeutic effectiveness while reducing side effects. For example, pharmacogenomic data can be analysed using algorithms to identify genetic markers that affect medication metabolism. This information can then be used to customise dose regimens based on the reactions of specific patients [52].

The integration of genetic data unique to each patient into molecular docking not only opens up new possibilities for unparalleled personalisation in drug design, but it also raises the possibility of safer and more effective treatments. The convergence of computational techniques and personalised medicine holds great promise for revolutionising precision medication design, enabling a paradigm shift towards customised treatments that take into account each patient's own genetic landscape [53].

4.1.2 Designing drugs based on individual molecular profiles

Molecular docking in conjunction with personalised medicine ushers in a new era of drug discovery wherein pharmaceutical interventions are finely customised according to individual molecular profiles. Using the abundance of data from omics technologies—genomics, transcriptomics, proteomics, and metabolomics, for example—this novel method creates medications that are precisely tailored to the distinctive features of each patient's molecular landscape. As a computational powerhouse, molecular docking plays a pivotal role in this process by forecasting the interactions between medications and particular biological targets in individual patients. Drugs that are designed for maximum efficacy and lowest side effects can be more easily designed because to docking algorithms, which analyse the complex aspects of an individual's genetic composition, gene expression patterns, and other molecular markers. Beyond the conventional one-size-fits-all approach, this customised medication design approach acknowledges the intrinsic differences in patients' molecular profiles that affect therapeutic reactions. The algorithms provide a personalised drug development roadmap by predicting the most successful drug-receptor interactions based on genetic markers, pathway activities, and other biological aspects [54].

The convergence of molecular docking and personalised medicine holds great promise for revolutionary developments in drug design. Individual molecular profiles are used to build customised medications that have the potential to dramatically enhance treatment outcomes, reduce side effects, and usher in a new era of precision medicine in which pharmaceutical interventions are as distinct as the persons they are intended to treat [55].

4.2 Target identification and validation

4.2.1 Utilising molecular docking in target selection

Target identification and validation are critical stages in the molecular docking and personalised medicine integration process, where molecular docking techniques are essential for the deliberate selection of therapeutic targets. In order to forecast and assess the binding interactions between possible drug candidates and a wide range of molecular targets, this procedure makes use of computational methods. Molecular docking facilitates the methodical investigation of the wide-ranging biological terrain by reducing the number of pertinent targets selected according to parameters like binding affinity and specificity. Docking algorithms use in silico simulations to forecast the possible interactions between a drug candidate and different molecular targets. This allows for the prediction of potential success in modifying specific pathways or biomolecules linked to a given disease. This strategy is particularly useful in the personalised medicine setting, where accurate target selection is required due to individual differences in genetic and molecular profiles. Molecular docking aids in the discovery of targets that are in line with the distinct features of each patient's illness phenotype by integrating patient-specific data [56].

In addition to streamlining the drug development process, target selection based on molecular docking guarantees that therapeutic interventions are precisely tailored to the unique molecular settings of individual patients. The combination of computational methods with personalised medicine holds the potential to improve drug development's effectiveness and efficiency while paving the way for more specialised and efficient therapies [57].

4.2.2 Validating targets for personalised therapies

Molecular docking approaches support the crucial stage of target discovery and validation in the personalised medicine-molecular docking integration, helping to rigorously validate targets for customised therapeutics. Molecular docking is a computer method that predicts and analyses the binding interactions between possible drug candidates and selected targets, assisting in the validation of the appropriateness of these targets. Molecular docking refines target selection based on binding affinity, specificity, and pharmacological significance by evaluating the viability of drug-receptor interactions using virtual screening and simulation analysis. In the context of personalised medicine, this procedure is especially helpful since it ensures that targets are validated and identified based on the distinct genetic and molecular traits of individual patients [58].

A thorough understanding of the ways in which particular molecular variants contribute to the aetiology of disease is necessary for the validation of targets for personalised therapeutics. This procedure is made easier by molecular docking, which predicts how well drug candidates would bind to targets linked to specific patient profiles and provides information about the therapeutic intervention's likelihood of success. Molecular docking and personalised medicine working together to validate targets not only expedites the conversion of genomic and omics data into workable treatment plans but also guarantees that customised treatments are supported by substantial scientific data. In the era of precision medicine, this confluence signifies a transformative approach where computational approaches are essential to the validation of targets, opening the door to more efficient and customised treatments [59].

4.3 Examples of personalised medicine success using molecular docking

A number of notable case studies highlight the effective fusion of molecular docking with personalised medicine, illustrating the revolutionary potential of this cooperative strategy in practical therapeutic contexts. A good example is the treatment of some cancers, where it is essential to find genetic alterations unique to each patient in order to customise treatment. Researchers can simulate the interactions between medications and altered proteins to anticipate which targeted therapies will work best by using molecular docking. Patients who have used this strategy have had amazing results, including higher response rates and less side effects [60].

Molecular docking-guided personalised therapy has shown promise in treating infectious illnesses like HIV. To create antiretroviral medications that precisely target distinct viral weaknesses, researchers use docking algorithms to account for individual variances in viral strains and host variables. Drug resistance has decreased and treatment efficacy has increased as a result of this customised approach. Furthermore, the combination of molecular docking and personalised medicine has demonstrated promise in the identification of medications that particularly target individual genetic and molecular variables leading to neurodegenerative illnesses like Alzheimer's disease. These case studies highlight the observable benefits of molecular docking and personalised medicine, demonstrating how this combined approach is changing the face of healthcare by providing tailored, efficient, and focused treatments for a range of illnesses [61].

4.3.1 Impact on patient outcomes

Molecular docking in conjunction with personalised therapy has shown impressive effects on patient outcomes in a variety of medical contexts. Molecular docking simulations have been used to identify individual genetic profiles, which have been used to adapt cancer treatments in oncology. This approach has resulted in considerable improvements in response rates and a decrease in side effects. Individualised therapy recipients benefit from increased treatment efficacy, longer survival times, and general improvements in their quality of life. Treatment techniques for infectious diseases have been revolutionised by personalised approaches informed by molecular docking, especially in the context of antiretroviral therapy for HIV. With better viral suppression, decreased resistance to therapy, and a greater chance of obtaining undetectable viral levels, the targeted targeting of patient-specific viral strains has improved long-term patient outcomes [62].

The possibility for slowing the progression of neurodegenerative illnesses has been demonstrated by the merging of molecular docking with personalised treatment. Patients with diseases such as Alzheimer's disease have reported improved functional outcomes and cognitive stabilisation when therapies are customised to each patient's unique genetic and molecular variables. Together, these case studies highlight the real influence on patient outcomes and highlight the revolutionary potential of molecular docking-driven personalised therapy. This comprehensive strategy is improving therapeutic precision and improving overall well-being and prognosis of patients receiving personalised interventions by tailoring treatments to each patient's unique genetic and molecular profile [63].

5. Challenges and future directions

5.1 Ethical considerations

5.1.1 Privacy concerns in genomic data

Growing developments in personalised medicine, particularly with regard to privacy in the handling of genomic information, are brought to light by the combination of genomic data and molecular docking. Concerns over possible misuse or unauthorised access to sensitive information are raised by genomic data since it is intrinsically personal and uniquely identifying. Risk of genetic data breaches and its possible effects on persons is one of the main challenges. Because genomic data is so rich, getting it without authorization could jeopardise someone's privacy and put them at risk of stigmatisation, discrimination, or improper use of their genetic information [64].

Furthermore, it is crucial to make sure that patient data is de-identified and anonymized because personalised medicine depends on the compilation and analysis of massive genetic databases. A difficult ethical conundrum arises when attempting to strike a balance between the value of shared data for research purposes and protecting individual privacy. Strong data governance frameworks, strict security measures, and open informed consent processes are necessary to address these issues. The future of personalised medicine will require continued communication between scientists, physicians, legislators, and the general public to develop moral guidelines that protect individuals' privacy and promote progress in the field of genetic medicine. As the field develops, handling genomic data in a proactive and moral manner will be essential to guaranteeing the ethical and responsible advancement of personalised medicine [65].

5.1.2 Informed consent and patient autonomy

One of the most important ethical issues in the rapidly developing fields of genetic data integration and personalised medicine is ensuring patient autonomy and obtaining informed permission. Patients must have a thorough comprehension of genomic information due to its complexity before consenting to engage in research or have genomic testing done on them. Maintaining ethical standards requires making sure people are properly educated about the possible consequences, hazards, and advantages of sharing their genetic information. There are questions about how genomic data is dynamic and how our understanding of genetics changes over time. Respecting patient autonomy requires continuing to communicate with patients, giving them information, and getting their consent again as knowledge expands [66].

Furthermore, the difficulty is striking a balance between personal freedom and the group advantages of combined genetic data. Finding the correct balance entails being open and honest about the possible effects on society of genetic information sharing for research, encouraging a feeling of shared responsibility while honouring personal preferences. The informed consent process needs to be improved, educational programmes need to be implemented, and technology needs to be embraced for dynamic consent models in the future. Giving patients a comprehensive knowledge of the consequences of sharing their genetic information helps to maintain moral principles, build confidence, and promote the ethical development of personalised medicine [67].

5.2 Technical challenges

5.2.1 Integration of large-scale omics data

In the field of personalised medicine, large-scale omics data integration presents formidable technical obstacles. A number of obstacles need to be overcome before genomic, transcriptomic, proteomic, and metabolomic data can be effectively used in patient treatment, given their increasing volume and complexity. Creating reliable computational frameworks that can handle a variety of omics datasets and extract insightful information is a major issue. Advanced bioinformatics tools and algorithms that can reconcile various data types, take variability into account, and identify pertinent patterns are necessary for the integration of multi-dimensional data [68].

The requirement for scalable and interoperable platforms that enable smooth integration across various healthcare systems and research institutes represents another technical challenge. By standardising data formats and protocols, personalised medicine projects can be expanded in scope and depth by facilitating the sharing, aggregation, and collective analysis of omics data. Real-time omics data integration into clinical processes also necessitates advancements in data storage, processing speed, and secure sharing protocols. In addition to improving personalised medicine's accuracy, resolving these technological issues will expedite the conversion of omics research results into practical knowledge for medical professionals. In order to pave the way for a more technically sound and clinically significant era in personalised healthcare, future directions in personalised medicine will necessitate concentrated efforts in creating state-of-the-art informatics infrastructure, encouraging collaboration between computational biologists and clinicians, and establishing standards for data sharing [69].

5.2.2 Validation of molecular docking predictions

Ensuring rigorous confirmation of computational predictions is a continuous technical problem in the integration of molecular docking and personalised medicine.

In silico drug design can be greatly enhanced by molecular docking, however there are challenges in guaranteeing the precision and dependability of its predictions. In molecular docking, validation is evaluating how well computational models forecast the binding interactions seen in real-world environments. Sophisticated methods like nuclear magnetic resonance spectroscopy or X-ray crystallography are frequently needed for experimental validation in order to ascertain the true structure of ligand-receptor complexes [55].

The complexity of biological systems and the wide range of variables affecting molecular interactions present a challenge. Complexity in validation is increased by variations in protein conformations, effects of solvents, and precise depiction of binding sites. Creating benchmark datasets, which use known ligand-receptor complexes as a foundation for assessing docking predictions, is necessary to address this difficulty. Molecular docking simulations must be continuously improved in order to increase the predicted accuracy of scoring functions and algorithms. Moving forward, the scientific community will need to adopt standardised protocols for validation, leverage machine learning techniques to increase docking accuracy, and advance experimental approaches for high-throughput validation. It is imperative to overcome these technical obstacles in order to validate the accuracy of molecular docking predictions and guarantee their application in trustworthy personalised medicine [50].

5.3 Future prospects

5.3.1 Advancements in technology and methodology

Personalised medicine has a bright future ahead of it thanks to expected advances in technology and practice. With the ongoing increase in processing power, new approaches and technologies have the potential to completely transform the industry. Developments in machine learning and artificial intelligence (AI) will be crucial. It is anticipated that AI algorithms would improve molecular docking prediction accuracy, allowing for more accurate drug-receptor interaction simulations. The utilisation of machine learning models is expected to aid in the discernment of complex patterns present in extensive omics datasets, hence expediting the identification of new therapeutic targets and biomarkers. Understanding cellular heterogeneity will change dramatically with the introduction of single-cell omics technology. With the help of this high-resolution method, illness states and treatment outcomes can be more accurately described and individual cell behaviour can be revealed in a way never possible before [70].

Moreover, the amalgamation of multi-omics data, encompassing transcriptomics, proteomics, metabolomics, and genomes, would provide a comprehensive perspective of a person's molecular terrain. Physicians will be able to create thorough profiles for individualised diagnosis and treatment plans thanks to platforms for advanced analytics and data integration. Technologies based on CRISPR-mediated gene editing show promise for accurate genetic interventions, enabling the correction of genetic abnormalities that underlie a variety of diseases. This groundbreaking skill creates opportunities for really customised therapeutic approaches. In conclusion, a revolutionary era where healthcare is progressively customised to each individual's own features is being fostered by the synergistic fusion of cutting-edge technologies and creative techniques that characterise the future of personalised medicine [71].

5.3.2 Potential breakthroughs in personalised medicine and molecular docking

Molecular docking and personalised medicine have enormous potential for significant advances that could completely change the way that healthcare is provided. The combination of quantum computing with molecular docking simulations holds promise for a breakthrough. The speed and accuracy of docking predictions could be greatly increased by the unmatched computational power of quantum computers, opening up new avenues for drug design. The accuracy of molecular docking predictions could be improved by higher-resolution insights into protein structures that can be obtained using structural biology techniques like cryo-electron microscopy. This may result in the more accurate determination of the best drug-receptor combinations, particularly for difficult targets. There is tremendous promise in the combination of CRISPR-based gene editing technology and personalised medicine [72].

Accurate genetic adjustments directed by molecular docking predictions may open the door to customised therapeutic interventions at the genomic level, tackling the underlying causes of hereditary illnesses. Advanced omics technologies provide promise for the identification of new molecular signatures in the field of biomarker discovery. The combination of machine learning techniques and multi-omics data may reveal patterns that were not previously identified, leading to the identification of reliable biomarkers for prognosis, therapy response, and illness diagnosis. With these developments, the combination of molecular docking and personalised medicine is set to usher in a new era of healthcare that not only customises treatment to each patient's unique molecular profile but also uses state-of-the-art technologies to precisely optimise therapeutic outcomes [73].

6. Conclusion

Pharmacological discovery is undergoing a radical metamorphosis at the dynamic nexus of molecular docking and personalised medicine, paving the way for a future in which healthcare is more individually customised. Molecular docking has become a key component in the logical design of personalised medicines thanks to the convergence of genetic knowledge, cutting-edge omics technology, and computational power. Personalised medicine is important because it has the potential to completely transform patient care as we negotiate this new terrain. Molecular docking's predictive capability combined with the capacity to interpret individual genetic variations allows clinicians to not only diagnose patients with unprecedented precision but also to tailor treatment strategies to each patient's unique molecular subtleties. This revolution is enhanced by molecular docking, a computational ally that acts as a link between theoretical predictions and experimental validations. It speeds up the process of finding new drugs by pointing scientists in the direction of the best possible drug-receptor interactions with amazing effectiveness. The combination of artificial intelligence, large-scale omics data, and structural biology advances increases the likelihood of molecular disease knowledge and targeting advances. The possibilities for the future are exciting: possible developments in biomarker identification, CRISPR-based gene editing, and quantum computing might completely change the field of personalised medicine. The idea of customising medication discovery for individual patients is becoming increasingly real as these technologies advance, offering a new paradigm in healthcare where interventions are proactive rather than reactive, taking into account each patient's own molecular composition. To sum up, the combination of molecular docking and personalised medicine is a bright light that points the way towards a future in which healthcare is highly predictive, genuinely personalised, and deeply affects each patient's quality of life.

Author details

Noopur Khare[1] and Pragati Khare[2]*

1 Bhai Gurdas Group of Institutions, Sangrur, Punjab, India

2 Department of Pharmacy, Bhagwant University, Ajmer, Rajasthan, India

*Address all correspondence to: pragatikhare10@gmail.com

References

[1] Moerenhout T, Devisch I, Cornelis GC. E-health beyond technology: Analyzing the paradigm shift that lies beneath. Medicine, Health Care and Philosophy. 2018;**21**:31-41

[2] Kunduru AR. Machine learning in drug discovery: A comprehensive analysis of applications, challenges, and future directions. International Journal on Orange Technologies. 2023;**5**(8):29-37

[3] Ginsburg GS, Willard HF. The foundations of genomic and personalized medicine. In: Essentials of Genomic and Personalized Medicine. Academic Press; 2010. pp. 1-10

[4] Khare N, Maheshwari SK, Rizvi SM, Albadrani HM, Alsagaby SA, Alturaiki W, et al. Homology modelling, molecular docking and molecular dynamics simulation studies of CALMH1 against secondary metabolites of bauhinia variegata to treat Alzheimer's disease. Brain Sciences. 2022;**12**(6):770

[5] Ahmed Z, Mohamed K, Zeeshan S, Dong X. Artificial intelligence with multi-functional machine learning platform development for better healthcare and precision medicine. Database. 2020;**2020**:baaa010

[6] Sharma G. The human genome project and its promise. Journal of Indian College of Cardiology. 2012;**2**(1):1-3

[7] Pirmohamed M. Personalized pharmacogenomics: Predicting efficacy and adverse drug reactions. Annual Review of Genomics and Human Genetics. 2014;**15**:349-370

[8] Nassar SF, Raddassi K, Ubhi B, Doktorski J, Abulaban A. Precision medicine: Steps along the road to combat human cancer. Cell. 2020;**9**(9):2056

[9] Redekop WK, Mladsi D. The faces of personalized medicine: A framework for understanding its meaning and scope. Value in Health. 2013;**16**(6):S4-S9

[10] Teama S. DNA polymorphisms: DNA-based molecular markers and their application in medicine. In: Genetic Diversity and Disease Susceptibility. 2018. pp. 25-32

[11] Kendler KS, Aggen SH, Czajkowski N, Røysamb E, Tambs K, Torgersen S, et al. The structure of genetic and environmental risk factors for DSM-IV personality disorders: A multivariate twin study. Archives of General Psychiatry. 2008;**65**(12):1438-1446

[12] Ozomaro U, Wahlestedt C, Nemeroff CB. Personalized medicine in psychiatry: Problems and promises. BMC Medicine. 2013;**11**(1):1-35

[13] Hasanzad M, Sarhangi N, Ehsani Chimeh S, Ayati N, Afzali M, Khatami F, et al. Precision medicine journey through omics approach. Journal of Diabetes & Metabolic Disorders. 2022;**21**(1):881-888

[14] Zhang J, Chiodini R, Badr A, Zhang G. The impact of next-generation sequencing on genomics. Journal of Genetics and Genomics. 2011;**38**(3):95-109

[15] LaFramboise T. Single nucleotide polymorphism arrays: A decade of biological, computational and technological advances. Nucleic Acids Research. 2009;**37**(13):4181-4193

[16] Weinshilboum RM, Wang L. Pharmacogenomics: Precision medicine

and drug response. In: Mayo Clinic Proceedings. Vol. 92, no. 11. Elsevier; 2017. pp. 1711-1722

[17] Berliner JL, Cummings SA, Boldt Burnett B, Ricker CN. Risk assessment and genetic counseling for hereditary breast and ovarian cancer syndromes—Practice resource of the National Society of genetic Counselors. Journal of Genetic Counseling. 2021;**30**(2):342-360

[18] Veenstra TD. Omics in systems biology: Current progress and future outlook. Proteomics. 2021;**21**(3-4):2000235

[19] Garg M. RNA sequencing: A revolutionary tool for transcriptomics. In: Advances in Animal Genomics. Academic Press; 2021. pp. 61-73

[20] Tsimberidou AM, Fountzilas E, Bleris L, Kurzrock R. Transcriptomics and solid tumors: The next frontier in precision cancer medicine. In: Seminars in Cancer Biology. Vol. 84. Academic Press; 2022. pp. 50-59

[21] Das PP, Rana S, Muthamilarasan M, Kannan M, Ghazi IA. Omics approaches for understanding plant Defense response. Omics Technologies for Sustainable Agriculture and Global Food Security. 2021;**1**:41-83

[22] Su M, Zhang Z, Zhou L, Han C, Huang C, Nice EC. Proteomics, personalized medicine and cancer. Cancers. 2021;**13**(11):2512

[23] Manzoni C, Kia DA, Vandrovcova J, Hardy J, Wood NW, Lewis PA, et al. Genome, transcriptome and proteome: The rise of omics data and their integration in biomedical sciences. Briefings in Bioinformatics. 2018;**19**(2):286-302

[24] Dai X, Shen L. Advances and trends in omics technology development. Frontiers in Medicine. 2022;**9**:911861

[25] Segers K, Declerck S, Mangelings D, Heyden YV, Eeckhaut AV. Analytical techniques for metabolomic studies: A review. Bioanalysis. 2019;**11**(24):2297-2318

[26] Singh S, Sarma DK, Verma V, Nagpal R, Kumar M. Unveiling the future of metabolic medicine: Omics technologies driving personalized solutions for precision treatment of metabolic disorders. Biochemical and Biophysical Research Communications. 2023;**682**(1):1-20

[27] Angione C. Human systems biology and metabolic modelling: A review—From disease metabolism to precision medicine. BioMed Research International. 2019:1-19

[28] Qoronfleh MW, Chouchane L, Mifsud B, Al Emadi M, Ismail S. THE FUTURE OF MEDICINE, healthcare innovation through precision medicine: Policy case study of Qatar. Life Sciences, Society and Policy. 2020;**16**(1):1-20

[29] La Cognata V, Morello G, Cavallaro S. Omics data and their integrative analysis to support stratified medicine in neurodegenerative diseases. International Journal of Molecular Sciences. 2021;**22**(9):4820

[30] Giordano S, Petrelli A. From single-to multi-target drugs in cancer therapy: When aspecificity becomes an advantage. Current Medicinal Chemistry. 2008;**15**(5):422-432

[31] Pokorska-Bocci A, Stewart A, Sagoo GS, Hall A, Kroese M, Burton H. 'Personalized medicine': what's in a name? Personalized Medicine. 2014;**11**(2):197-210

[32] Seymour CW, Gomez H, Chang CC, Clermont G, Kellum JA, Kennedy J, et al. Precision medicine for all? Challenges

and opportunities for a precision medicine approach to critical illness. Critical Care. 2017;**21**:1-1

[33] Mastrangelo A, Armitage G, E, García A, Barbas C. Metabolomics as a tool for drug discovery and personalised medicine. A review. Current Topics in Medicinal Chemistry. 2014;**14**(23):2627-2636

[34] Seyhan AA, Carini C. Are innovation and new technologies in precision medicine paving a new era in patients centric care? Journal of Translational Medicine. 2019;**17**:1-28

[35] Abdolmaleki A, Ghasemi JB, Ghasemi F. Computer aided drug design for multi-target drug design: SAR/QSAR, molecular docking and pharmacophore methods. Current Drug Targets. 2017;**18**(5):556-575

[36] Sivakumar KC, Haixiao J, Naman CB, Sajeevan TP. Prospects of multitarget drug designing strategies by linking molecular docking and molecular dynamics to explore the protein–ligand recognition process. Drug Development Research. 2020;**81**(6):685-699

[37] Adelusi TI, Oyedele AQ, Boyenle ID, Ogunlana AT, Adeyemi RO, Ukachi CD, et al. Molecular modeling in drug discovery. Informatics in Medicine Unlocked. 2022;**29**:100880

[38] Khare N, Maheshwari SK, Jha AK. Screening and identification of secondary metabolites in the bark of Bauhinia variegata to treat Alzheimer's disease by using molecular docking and molecular dynamics simulations. Journal of Biomolecular Structure and Dynamics. 2021;**39**(16):5988-5998

[39] Parween A, Singh PK, Anamika Y, Khare N, Rai NP, Bajpai M, et al. Molecular docking of quinolone against INHA to treat tuberculosis. International Journal for Research in Applied Science and Engineering Technology. 2020;**8**(6):2421-2427

[40] Naqvi AA, Mohammad T, Hasan GM, Hassan MI. Advancements in docking and molecular dynamics simulations towards ligand-receptor interactions and structure-function relationships. Current Topics in Medicinal Chemistry. 2018;**18**(20):1755-1768

[41] Ramírez D. Computational methods applied to rational drug design. The Open Medicinal Chemistry Journal. 2016;**10**:7

[42] Mohan A, Krishnamoorthy S, Sabanayagam R, Schwenk G, Feng E, Ji HF, et al. Pharmacophore based virtual screening for identification of effective inhibitors to combat HPV 16 E6 driven cervical cancer. European Journal of Pharmacology. 2023;**957**:175961

[43] Vázquez J, López M, Gibert E, Herrero E, Luque FJ. Merging ligand-based and structure-based methods in drug discovery: An overview of combined virtual screening approaches. Molecules. 2020;**25**(20):4723

[44] Lee S, Barron MG. Development of 3D-QSAR model for acetylcholinesterase inhibitors using a combination of fingerprint, molecular docking, and structure-based pharmacophore approaches. Toxicological Sciences. 2015;**148**(1):60-70

[45] Ballante F, Kooistra AJ, Kampen S, de Graaf C, Carlsson J. Structure-based virtual screening for ligands of G protein–coupled receptors: What can molecular docking do for you? Pharmacological Reviews. 2021;**73**(4):1698-1736

[46] Di Filippo JI, Cavasotto CN. Guided structure-based ligand identification and design via artificial intelligence modeling. Expert Opinion on Drug Discovery. 2022;**17**(1):71-78

[47] Lionta E, Spyrou G, Vassilatis K, D, Cournia Z. Structure-based virtual screening for drug discovery: Principles, applications and recent advances. Current Topics in Medicinal Chemistry. 2014;**14**(16):1923-1938

[48] Sethi A, Joshi K, Sasikala K, Alvala M. Molecular docking in modern drug discovery: Principles and recent applications. Drug Discovery and Development-New Advances. 2019;**2**:1-21

[49] Wang T, Wu MB, Lin JP, Yang LR. Quantitative structure–activity relationship: Promising advances in drug discovery platforms. Expert Opinion on Drug Discovery. 2015;**10**(12):1283-1300

[50] Saikia S, Bordoloi M. Molecular docking: Challenges, advances and its use in drug discovery perspective. Current Drug Targets. 2019;**20**(5):501-521

[51] Aouidate A, Ghaleb A, Ghamali M, Chtita S, Ousaa A, Choukrad MB, et al. Furanone derivatives as new inhibitors of CDC7 kinase: Development of structure activity relationship model using 3D QSAR, molecular docking, and in silico ADMET. Structural Chemistry. 2018;**29**:1031-1043

[52] Sneha P, Doss CG. Molecular dynamics: New frontier in personalized medicine. Advances in Protein Chemistry and Structural Biology. 2016;**102**:181-224

[53] Gayathiri E, Prakash P, Kumaravel P, Jayaprakash J, Ragunathan MG, Sankar S, et al. Computational approaches for modeling and structural design of biological systems: A comprehensive review. Progress in Biophysics and Molecular Biology. 2023;**285**(1):17-32

[54] Waheed S, Li Z, Zhang F, Chiarini A, Armato U, Wu J. Engineering nano-drug biointerface to overcome biological barriers toward precision drug delivery. Journal of Nanobiotechnology. 2022;**20**(1):395

[55] Moingeon P, Kuenemann M, Guedj M. Artificial intelligence-enhanced drug design and development: Toward a computational precision medicine. Drug Discovery Today. 2022;**27**(1):215-222

[56] Agu PC, Afiukwa CA, Orji OU, Ezeh EM, Ofoke IH, Ogbu CO, et al. Molecular docking as a tool for the discovery of molecular targets of nutraceuticals in diseases management. Scientific Reports. 2023;**13**(1):13398

[57] Ivasiv V, Albertini C, Gonçalves AE, Rossi M, Bolognesi ML. Molecular hybridization as a tool for designing multitarget drug candidates for complex diseases. Current Topics in Medicinal Chemistry. 2019;**19**(19):1694-1711

[58] Özenver N, Efferth T. Integration of Phytochemicals and Phytotherapy into Cancer Precision Medicine. Approaching Complex Diseases: Network-Based Pharmacology and Systems Approach in Bio-Medicine; 2020. pp. 355-392

[59] Hopkins CE, Brock T, Caulfield TR, Bainbridge M. Phenotypic screening models for rapid diagnosis of genetic variants and discovery of personalized therapeutics. Molecular Aspects of Medicine. 2023;**91**:101153

[60] Hassan M, Awan FM, Naz A, deAndrés-Galiana EJ, Alvarez O, Cernea A, et al. Innovations in genomics and big data analytics for personalized medicine and health care: A review. International Journal of Molecular Sciences. 2022;**23**(9):4645

[61] Melge AR, Kumar LG, Pavithran K, Nair SV, Manzoor K. Predictive models for designing potent tyrosine kinase inhibitors in chronic myeloid leukemia for understanding its molecular mechanism of resistance by molecular docking and dynamics simulations. Journal of Biomolecular Structure and Dynamics. 2019;**37**(18):4747-4766

[62] Husnain A, Rasool S, Saeed A, Hussain HK. Revolutionizing pharmaceutical research: Harnessing machine learning for a paradigm shift in drug discovery. International Journal of Multidisciplinary Sciences and Arts. 2023;**2**(2):149-157

[63] Gnanaraj C, Sekar M, Fuloria S, Swain SS, Gan SH, Chidambaram K, et al. In silico molecular docking analysis of karanjin against alzheimer's and parkinson's diseases as a potential natural lead molecule for new drug design, development and therapy. Molecules. 2022;**27**(9):2834

[64] Najafi A, Emami N, Samad-Soltani T. Integration of genomics data and electronic health records toward personalized medicine: A targeted review. Frontiers in Health Informatics. 2021;**10**(1):86

[65] Hulsen T, Jamuar SS, Moody AR, Karnes JH, Varga O, Hedensted S, et al. From big data to precision medicine. Frontiers in Medicine. 2019;**6**:34

[66] Ormond KE, Cho MK. Translating personalized medicine using new genetic technologies in clinical practice: The ethical issues. Personalized Medicine. 2014;**11**(2):211-222

[67] Bruynseels K, Santoni de Sio F, Van den Hoven J. Digital twins in health care: Ethical implications of an emerging engineering paradigm. Frontiers in Genetics. 2018;**9**:31

[68] Grapov D, Fahrmann J, Wanichthanarak K, Khoomrung S. Rise of deep learning for genomic, proteomic, and metabolomic data integration in precision medicine. Omics: A Journal of Integrative Biology. 2018;**22**(10):630-636

[69] Martinez-Garcia M, Hernández-Lemus E. Data integration challenges for machine learning in precision medicine. Frontiers in Medicine. 2022;**8**:784455

[70] Arabi AA. Artificial intelligence in drug design: Algorithms, applications, challenges and ethics. Future. Drug Discovery. 2021;**3**(2):FDD59

[71] Bhattacharjee G, Gohil N, Khambhati K, Mani I, Maurya R, Karapurkar JK, et al. Current approaches in CRISPR-Cas9 mediated gene editing for biomedical and therapeutic applications. Journal of Controlled Release. 2022;**343**:703-723

[72] Auffray C, Chen Z, Hood L. Systems medicine: The future of medical genomics and healthcare. Genome Medicine. 2009;**1**:1-1

[73] Strianese O, Rizzo F, Ciccarelli M, Galasso G, D'Agostino Y, Salvati A, et al. Precision and personalized medicine: How genomic approach improves the management of cardiovascular and neurodegenerative disease. Genes. 2020;**11**(7):747

Chapter 7

Molecular Docking: An Insight from Drug Discovery to Drug Repurposing Approach

Sana Shamim, Rabya Munawar, Yasmeen Rashid, Sheikh Muhammad Zesshan Qadar, Rabia Bushra, Irshad Begum, Muhammad Imran and Tehseen Quds

Abstract

The impact of computer-aided drug designing in the field of medicinal chemistry has created a boon in the drug discovery process. Molecular docking is an integral part of bioinformatics that deals with protein-ligand interactions, binding conformations, and affinity predictions. It has shown to be a rapid, easy, and affordable method in business and research settings alike. The advancement in the hardware and software has led to enhanced molecular dynamic simulations and thus authenticate the computational results. This has created a great impact in minimizing the cost and time involved in the drug discovery process. It has also helped in identifying the rationale for drug repurposing approaches. This chapter will give in-depth knowledge of the importance of molecular docking in drug designing and discovery, their impact on drug repurposing, and success stories of the in silico approach in drug discovery and repurposing.

Keywords: drug design, drug discovery, molecular docking, drug repurposing, protein-ligand interactions

1. Introduction

As estimated by a study in 2016 by the Tufts Center for the Drug Development Study, the cost of a new drug development till the marketing step has increased by almost 144%, which sometimes does not go well if the drug gets recalled [1]. De novo drug development is a difficult, time-consuming, and costly procedure that needs a significant financial commitment. It can take anywhere from 13 to 15 years and 0.8 to 3 billion dollars on average to develop a medicine from a theoretical hypothesis to a new commercial medicinal molecule [2–5].

Drug research and discovery typically entail four steps of preclinical testing and clinical trials. According to estimates, the Food and Drug Administration (FDA) evaluates around nine out of ten drugs each year. Only 10% of the medications enrolled in phase 1 clinical trials are authorized; the remaining treatments fail because of severe toxicity or ineffectiveness. The primary cause of these attritions is incorrect medication target or response identification. Therefore, searching for the correct lead compound is crucial to the project's overall success in the difficult process of drug discovery. This not only contributes to reducing the economic burden of society but also manages the timeline associated with dealing with the morbidity or mortality of the disease as evident from the COVID-19 scenario.

2. Drug designing and discovery

Drug development is a long, arduous process that costs the galaxy. Preclinical testing is typically the first step in the four-phase clinical evaluation process that follows drug discovery and development. It has historically changed over the ages in step with advances in science and technology. Historically, the drug development approaches consist of *Serendipity* where drugs' original activities were found accidentally, as was the case of penicillin discovery [5]. A *synthetic and semi-synthetic approach* is related to the advancement in the field of chemistry. Replacing the natural origin compounds with synthetic or semi-synthetic alternatives reduces the cost of the product and meets the needs of the population. *Off-label use* means that they were being used for purposes different than those for which they were initially licensed [6]. For instance, propranolol was initially approved in the US in 1968 for the treatment of arrhythmias; but, 10 years later, it was also approved for the treatment of angina pectoris and antihypertensive medication [3].

Consequently, the goal of drug development is to create pharmaceuticals that bind to a certain protein target more firmly than the natural substrate. This allows for the modification or inhibition of the biological reaction that the target molecule catalyzes. Typically, drugs are found by accident through a process of trial and error utilizing high-throughput screening techniques that test the activity of

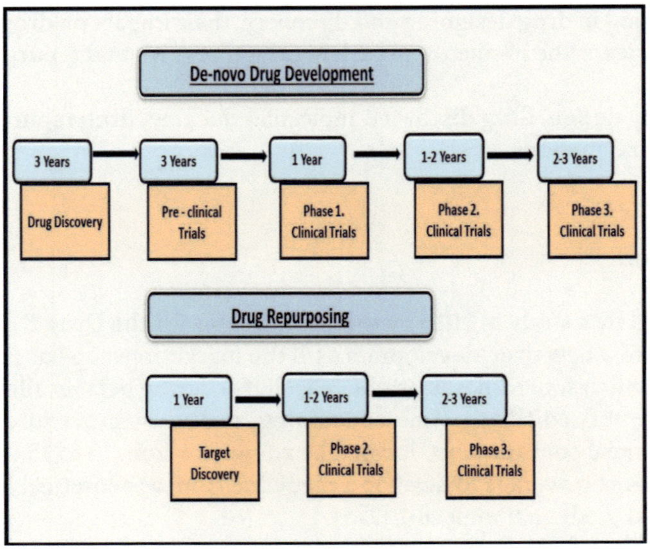

Figure 1.
Conceptual diagram of de novo drug development and drug repurposing. Drug repurposing starts with target discovery for an already-approved medication, which is immediately followed by phase 2 and 3 clinical trials. Phase 1 clinical studies and animal studies were not carried out because the information from these studies is already accessible for an already-approved medication [4, 7].

several compounds against a target in vitro. This is a costly and time-consuming process [7, 8].

Simulated molecular docking can be a helpful technique in the drug-discovery process if the target's 3D structure is known. By virtually screening compound databases, this in silico method enables the identification of promising therapeutic candidates more quickly and affordably. Once the drug candidates found by the virtual screening procedure have been discovered, additional research can be done on them through toxicological testing, clinical trials, lab experiments (synthesis), and other means **Figure 1** [9].

3. Drug repurposing strategies

Considering these challenges, it can be profitable to look for new uses and targets for pharmaceuticals that are already on the market. This strategy, known as "drug repurposing" (also known as "drug repositioning" or "drug re-profiling"), was initially presented by Ashburn and Thor in 2004 to find treatments for diseases for which there are none. In technical words, the way to identify the novel medication indications has already been approved. Such circumstances can include uncommon illnesses or a pandemic that is spreading quickly and for which there is no treatment [3–5]. There are two general repurposing approaches illustrated in **Figure 2**.

3.1 Knowledge-based repurposing

With this repurposing approach, models are created to forecast unidentified targets, biomarkers, or disease mechanisms by using drug-related data, such as drug targets, chemical structures, pathways, and side effects [10]. *Target-based* entails in silico screening of pharmacological compounds from drug libraries (*Drug bank,*

Figure 2.
Schematic illustration of repurposing approaches.

PubChem, Zinc database, ChemBl, etc.), for ligand-based screening or docking, after high-content screening and/or high-throughput (HCS/HTS) of medicinal compounds given proteins or biomarkers of interest [11, 12]. The target-based strategy has the benefit of screening almost all therapeutic molecules with known chemical structures. Pathway-based medication repurposing makes use of data from protein-interaction networks, metabolic pathways, and signaling pathways to forecast the relationship or resemblance between a drug and a condition (*KEGG, MetaCyc, FragariaCyc, etc.*) and *target mechanism-based drug repurposing*. Target mechanism-based drug repurposing integrates treatment omics data, protein interaction networks, and signaling pathway information to discover new mechanisms of action for drugs (*GenBank, UniProt, SWISS-2D PAGE, NCBI databases, etc.*) [13]. These repurposing approaches have the benefit of seeking to understand the mechanisms pertaining not only to pharmaceuticals or diseases but also to medication therapies for disorders.

3.2 Signature-based repurposing

Gene signature data from disease omics data are utilized in signature-based repurposing to identify novel off-targets or disease processes. To identify inverse relationships between drugs and diseases, this strategy examines the gene expression profiles of the medication and the illness. These methods have the benefit of revealing novel pharmacological mechanisms of action. Additionally, these strategies entail more molecular and/or genetic-level mechanisms than knowledge-based methods do [4, 14].

3.3 Phenotype-based repurposing

The availability of phenotypic data has opened new avenues for medication repositioning. System techniques have been using these kind of data more and more in the past few years to identify genetic features linked to human illnesses (*Phytozome, PubMed, Gramene database, EnsemblPlants, etc.*) [15].

4. Molecular docking studies

Molecular docking techniques play a major role in the design and development of novel medications [16–18] by predicting the experimental binding mechanism, affinity, and location of a small molecule (ligand) within the binding site of the target receptor (macromolecule) [19, 20]. The docking process (**Figure 3**) can accurately predict the native ligand posture at the receptor binding site (i.e., the experimental ligand geometry within a given tolerance limit) and the corresponding physical-chemical molecular interactions. It must also be able to accurately rank these ligands among the top compounds in the database and reliably distinguish binding from nonbinding molecules while analyzing large compound libraries [21, 22].

The primary step in the docking process is agreeable spatial and energetic matching between the ligand and receptor to obtain the optimal conformation. Furthermore, when the ligand attaches to the receptor, the structure of the binding site changes continually until a stable connection is correctly created. Only

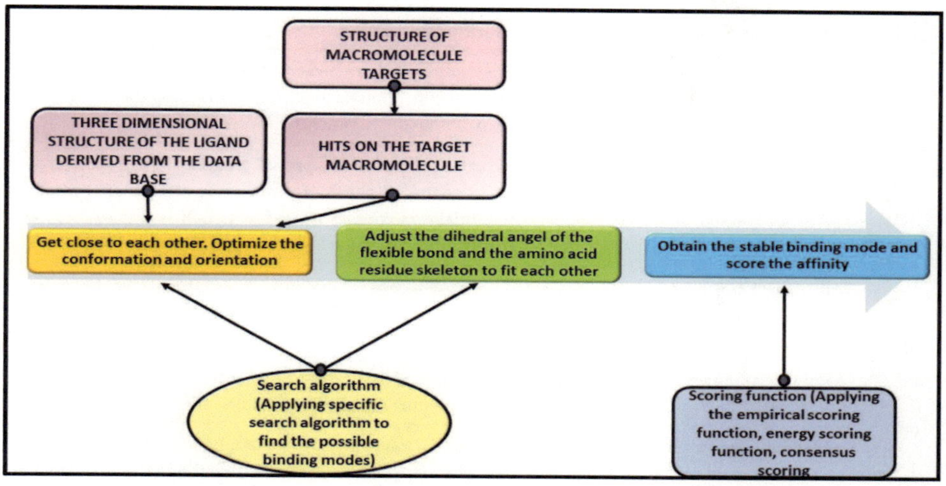

Figure 3.
Molecular docking process.

then can the configuration and charge distribution of the binding system be determined [16].

Pre-organization and geometric complementarity-based energy complementarity ensure that ligands and receptors have the most stable structures while consuming the least amount of free energy. The molecular docking program can help us find the optimal shape and orientation in terms of complementarity and pre-organization by employing a specific technique. Next, we may predict the binding affinity and look at the interactive mode using a scoring formula. Molecular docking is a widely utilized technique in the computer-aided structure-based drug development era to ascertain the anticipated geometry of a protein-ligand interaction [9].

Automation, parallelization, flexibility, sampling, optimizations, scoring, and *posture prediction* are all part of the docking process. Improvements over widely used algorithms are almost usually claimed: quicker docking, more precise affinity prediction, or more engaging graphics [21, 23]. A search algorithm and an energy scoring function are the basic tools of a docking methodology for generating and evaluating the ligand conformations [21]. Various molecular docking algorithms can be employed to predict protein-ligand configurations and order them based on the scoring functions integrated into every distinct docking approach [23, 24]. For molecular docking, the most popular heuristic search methods have been evolutionary, tabu, and simulated annealing [25].

To ascertain the most advantageous ligand conformation when bound to the target, docking techniques typically utilize an energy-based scoring system. A variety of scoring functions are available, including empirical, knowledge-based, force field-based, clustering and entropy-based, and consensus scoring approaches. The presence of water molecules at the active site is another aspect of docking target flexibility. It is generally agreed that lower energy scores represent better protein-ligand bindings than higher energy values. As a result, molecular docking can be formulated as an optimization problem where the objective is to determine the lowest energy ligand-binding mode **Table 1** [22, 25].

S.No.	Molecular docking approaches	Working principle
1	Fragment-based method	Fragment-based methods include breaking down the ligand into separate photons or particles, attaching the fragments, and then joining the fragments.
2	Inverse docking	When combined with a precise pharmacokinetic property, knowledge of all these targets can help assess a drug candidate's potential for toxicities and side effects. To conduct docking studies on a particular ligand, a special protocol is selected.
3	Blind Docking	This method was created to scan the entire target molecule interface to find putative peptide ligand binding sites and mechanisms of action.
4	Point complimentarily approach	These methods concentrate on contrasting the compositions and/or geometries of various molecules.
5	Matching approach	It places a strong emphasis on redundancy and determines the best place for the ligand atom within the site, which may lead to a ligand-receptor combination that still needs work.
6	Ligand fit approach	It is rapid and accurate method for docking tiny ligands into protein active sites while accounting for shape complementarity is called ligand fit.
7	Monte carlo approach	It causes a ligand's translations, rotations, and conformation to change randomly within an active site. An initial configuration value is assigned by it. After that, it evolves and generates a fresh configuration. The Metropolis criterion is used to determine whether the new configuration is maintained.
8	Distance geometry	It is possible to express several types of sequence properties in terms of intra- or intermolecular dimensions. These distances can be assembled and three-dimensional structures that work with them can be computed according to the framework of distance geometry.

Table 1.
Different molecular docking approaches [9, 23, 26].

5. Models of molecular docking

Emil Fischer's lock-and-key assumption in 1894 [8, 16, 27] states that both the ligand and the receptor may be thought of as rigid bodies and that their affinity is directly proportional to a geometric fit between their shapes, served as the foundation for the earliest known docking techniques. Subsequently, Koshland's "induced-fit" idea advocated that the ligand and receptor should be viewed as flexible during docking [8, 16, 28].

Unlike relatively autonomous side chains, each movement of the backbone affects numerous side chains. As a result, compared to flexible docking with a stiff receptor, the sampling process in a fully flexible receptor/ligand docking has a higher order of magnitude of degrees of freedom. As a result, these flexible docking algorithms can anticipate a molecule's binding mode and binding affinity concerning other compounds with greater accuracy than rigid body algorithms. The conformation ensemble model works beyond small induced-fit alterations, and proteins have been found to experience much larger conformational changes. A novel idea says that proteins are made from an ensemble of conformational states that already exist. Because of its flexibility, the protein can change states [26].

6. Types of docking

A technique for creating, characterizing, and altering compound topologies and interactions as well as the properties of these molecules that depend on their three-dimensional geometries is molecular modeling [26]. During the last 20 years, the development of molecular simulation (MS) technology has exhibited significant progress, especially since the 2013 Nobel Prize in Chemistry was awarded. As of right now, free energy calculation, molecular docking, homology modeling, and molecular dynamics simulation are all included in MS technology. The most popular approach among these in molecular modeling research is molecular docking [16, 29–33]. *Generally molecular docking can be Automated or manual.* The ligand is matched with its complementary group in the binding site during manual docking, where binding groups on both the ligand and the binding site are known. Each possible interaction's bonding distance is specified. To find the optimal fit as specified by the operator, the program shifts the molecule about inside the binding site. The matched groups are fitted so that they are within the desired bonding distances of one another rather than being directly superimposed.

It is possible to perform automatic docking, in which the software determines how to dock the ligand. The task for the docking program is twofold. I: it must arrange the ligand in various binding modes or orientations within the active site. II: to determine which binding modes are optimal, it must score each one. The order of complexity may be (a) target and ligand as rigid bodies; (b) ligand as a flexible body and target as a rigid substance; and (c) both target and ligand as flexible body.

The goals of docking simulations require, there are several types (**Figure 4**) of molecular docking processes that involve either flexible or stiff ligand/target combinations [21, 34, 35]. In *Rigid docking,* we assume that the compounds are stiff, and thus want to find a strategy to rearrange one of the compounds in three dimensions so that, in the context of a scoring system, it best fits (**Figure 5**) [23, 28, 29, 36]. It is a good fit for large systems like protein-protein and protein-nucleic acid. It is the simplest approach as it does not need several calculations [37, 38].

The flexible *docking approach* facilitates easy conformation modification of the docking system (receptor and ligand) during the process. As the variables of the receptor and ligand rise with the number of atoms, several other parameters must be considered. These include the computation's significant complexity and the docking procedure's excessive complexity. To effectively analyze the interaction between molecules, flexible docking is typically utilized [16, 39]. We assess molecular flexibility

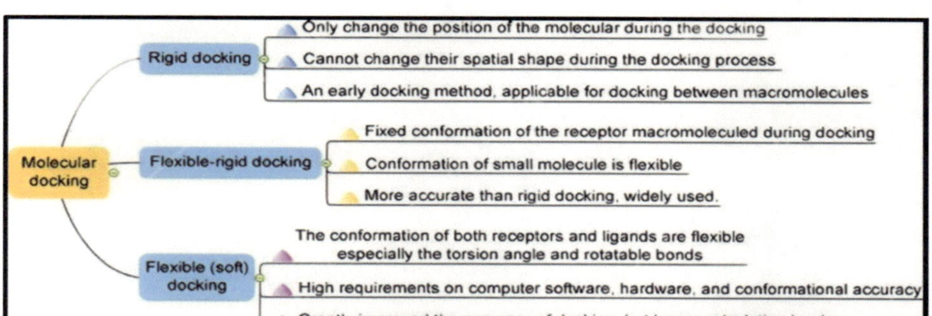

Figure 4.
Flow diagram of molecular docking types [8].

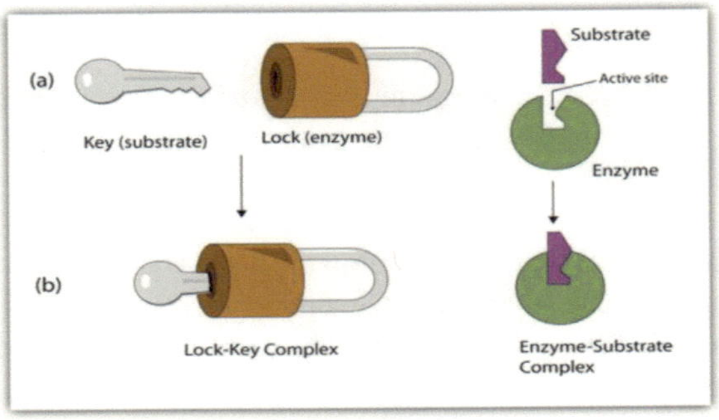

Figure 5.
Lock and key model [26].

in conjunction with transformation to find confirmations for the ligand and receptor molecules as they are present in the complex (**Figure 6**) [23, 26, 36, 40].

Another approach is *semi-flexible docking* where the ligand's conformation changes during the semi-flexible docking process, but the fixed receptor's conformation stays the same. Consequently, the noncritical portion of the ligand structure's bond length or angle can be fixed [41]. For the docking of macromolecules like proteins or nucleic acids as well as tiny ligand molecules, semi-flexible docking is appropriate [42]. It can

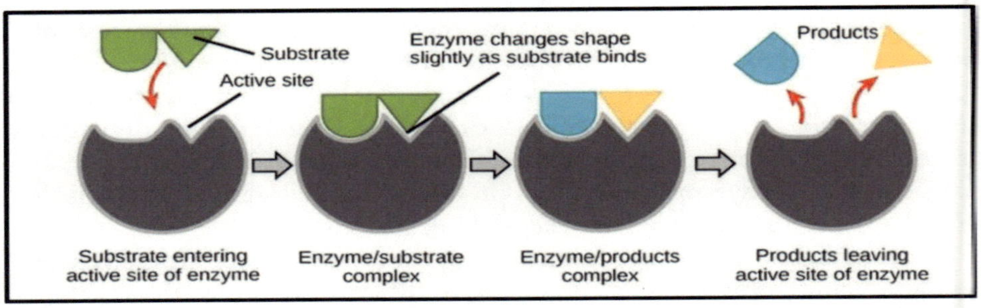

Figure 6.
Induced-fit model [26].

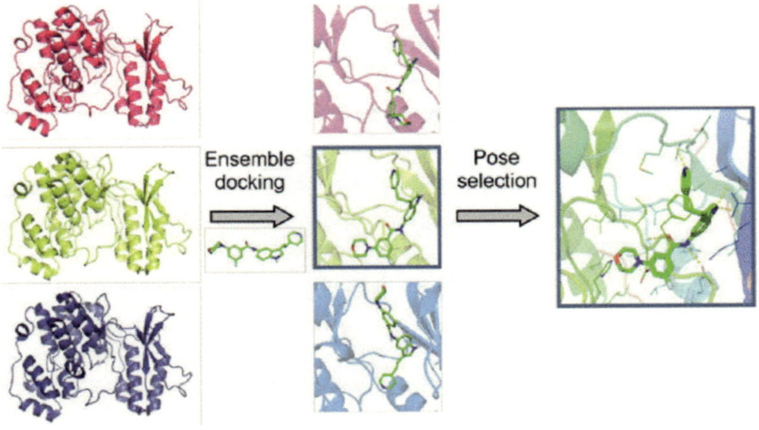

Figure 7.
Ensemble docking [21].

be used in a wider range of applications and considers the impact of ligand structural modifications [16]. *Ensemble docking* is more specific for protein interactions which clarifies the intricacy and flexibility of protein structural states. An ensemble of protein structures is used to bind with a ligand. Ensemble docking (**Figure 7**) is a potent technique for discretely representing the flexibility of a target receptor by docking molecules into several or multiple conformations of the receptor. This method offers a structural degree of freedom, where any ligand may find a compatible receptor, imitating the dynamic activity of the protein. In actuality, the latter is dependent upon the conformational space that the ensemble occupies. The general assumption is that ensemble docking is better than docking into a single receptor conformation [23, 40, 43, 44].

7. Methods in molecular docking

Molecular docking turned out to be a powerful tool in drug discovery. The first molecular docking algorithm was introduced in the 1980s and later methods in molecular docking have evolved drastically. Now, it is possible to analyze important molecular events such as binding conformations of the ligand and its interactions with the receptor that alleviate the ligand-receptor complex. Modern molecular docking algorithms exhibit the potential to quantitatively envisage binding energetics to categorize the docked ligands based on the binding energies of their complexes with the receptors. This section will introduce different existing molecular docking methods.

7.1 Conformational search methods

The basic goal of conformational search methods is to recognize the energetically favorable binding conformations of the ligand-receptor complex. Conformational search methods tend to incrementally amend ligand's torsional, translational, and rotational degrees of freedom by using either systemic or stochastic techniques.

7.1.1 Systematic methods

Systemic methods tend to perform an exhaustive and ordered search for the conformational landscape of the ligand by systematically varying its degrees of freedom. This method identifies all the combinations of structural constraints. Depending on the degrees of freedom of a particular ligand, several combinations are made that leads to a phenomenon called combinatorial explosion.

Surflex-Dock, Dock, FlexX, SLIDE, GLIDE, FRED, ADAM, EUDOC, Ehits, Flog, and Hammerhead are examples of Systematic methods [17].

7.1.2 Stochastic\random methods

Genetic algorithm (GA) is an exciting application of stochastic search algorithm where conformational optimization is carried out through the process of natural selection and genetics. It is used to find out a proper ligand's conformational space along with its energetically favorable binding modes. This algorithm is mainly used while working with flexible ligands or complex binding pockets [17].

Genetic algorithm begins with a preliminary population of different ligand conformers which are randomly generated. Using a scoring function that usually considers certain parameters like hydrogen bonds, hydrophobic contacts, salt bridges and steric hindrance, the fitness of all the generated ligand conformers is assessed. Based on the principle of "survival of the fittest," ligand conformers (also known as parent ligand conformers) showing better fitness scores are selected for the next round, that is, recombination and mutation. Whereas, torsional angles and other conformational parameters from two parent conformers are merged followed by their modest alterations in order to identify new regions of conformational space. This new population then enters another round of recombination and mutation to generate more improved ligand conformers. This process continues until the fulfillment of the required criterion where an acceptable fitness level is obtained.

Autodock, Gold, LigandFit, Molegro Virtual Docker, PLANTS, GlamDock, and MOE_Dock are examples of stochastic methods [17].

7.2 Energy scoring functions

To approximate the binding energies of the predicted ligand-receptor complex, certain scoring functions are used. These scoring functions normally rely on a few physicochemical parameters including intermolecular interactions, entropy, and desolvation [45]. There is an indirect relationship between the number of physicochemical parameters evaluated and the productivity of the docking algorithm because the computational cost increases with an increase in number of variables. An ideal scoring function should maintain a balance between score and speed.

Different scoring functions have been categorized into three different classes including force-field-based, empirical, and knowledge-based scoring functions [46].

7.2.1 Force-field-scoring function

This de novo method is applied to estimate the binding energy by generally adding the contributions of bonded and non-bonded terms [47]. However, due to the lack of realistic physical models for describing entropic effects and desolvation, the usage of this scoring function is limited [48]. Dock, AutoDock, and Molegro Virtual docker are well-known docking methods that utilize this scoring function.

7.2.2 Empirical scoring functions

These functions are derived from empirical observations and experimental data involving ligand-protein complexes with known binding energies rather than from first principles or physical equations. Due to the simplified way of estimating binding energies, empirical scoring functions are faster than force-field-based scoring functions. However, their dependence on the precision of data used to develop the training model is a major limitation. AutoDock, ChemScore, Surflex, and GlideScore are popular molecular docking method that utilizes empirical scoring functions [49, 50].

7.2.3 Knowledge-based scoring function

In knowledge-based scoring functions, statistical information retrieved from experimental structures is leveraged for examining the possibility of molecular interactions. Additionally, these scoring functions also consider the preference and occurrence of certain molecular interactions in the experimental data.

As these scoring functions neither calculate the binding energies de novo like fore-field scoring nor reproduce binding energies like empirical scoring, they present a balance between accuracy and speed. Molecular docking software that uses these scoring functions are SMoG, RF_Score, DrugScore, etc.

7.3 Covalent docking

Typically, drug designing protocols tend to search non-covalent ligands/inhibitors to avoid the toxicity problems of covalent inhibitors. However, during the SARS COVID-19 outbreak, the role of covalent drugs has been appreciated in terms of better efficacy compared to the native substrate of their targets.

Covalent docking is used to genuinely model covalent interactions between the ligands and receptors as it creates a tight bond between the electrophile (ligand) and nucleophile (protein), and it can produce chemical probes with high levels of potency and selectivity. This procedure offers the investigation of irreversible binding events that lead to the analysis of the crucial covalent drug interactions. This type of docking is mainly important in designing targeted therapies where irreversible ligand binding is required for improved efficacy and selectivity. However, the designing of the covalent ligand is somewhat difficult as certain molecular events including the creation and breaking of covalent bonds along with their rearrangements must be handled with Quantum Mechanics (QM). Notably, QM cannot adequately be operated by the traditional force fields or empirical techniques of molecular docking [51]; however, the requirement of QM can be avoided using faster and uncomplicated molecular modeling techniques. Fewer molecular docking methods have tackled the problem of covalent bond modeling including DOCK, GOLD, and AutoDock [17]. Numerous FDA-approved medications, including aspirin, warfarin, azacitidine, isoniazid, and others, have been discovered to exhibit covalent bonding [21].

7.4 Docking considering water molecule

Ligand-receptor interactions are simulated within the living system; certain docking algorithms have been introduced recently that evaluate the involvement of water molecules along with the ligand within the active (binding) site. The objective of considering the solvent setting improves the prediction accuracy, mostly in cases where water molecules play an important role in stabilizing the ligand-receptor complex.

In the Protein Data Bank, the crystal structures of more than 59% of the protein-ligand complexes were shown to contain at least one water molecule in their active site forming multiple hydrogen bonds between the ligand and the active site of the protein. These active site water molecules can be dislodged with the help of synthetic ligands or taken as a portion of the receptor protein. Certain methods are available to calculate the binding free energy for a given water molecule and, thus, distinguish between the dislocatable and structural water molecules [52].

Monte Carlo simulations, GOLD, AutoDock, HINT force field, etc., can perform molecular docking investigations considering structural water [53]. Nonetheless, there is a noticeable disparity in accuracy and capability across the many docking software versions that are currently on the market. In the literature, around 59 molecular docking tools have been published over the past 20 years for usage in academia and business, including numerous others.

8. File formats used in molecular docking

Different molecular docking programs rely on specific file formats of protein receptors and ligands. These file formats offer a standardized way to represent molecular structures of ligands and receptor proteins, facilitating the harmonization of different molecular docking software. The selection of a specific file format often depends on the specific requirements of the docking software being used. Some of the important file formats used in molecular docking are described here.

8.1 MOL2 (Tripos Mol2)

A Tripos Mol2 file (. mol2) is an ASCII file that is too absolute and manageable to represent a SYBYL molecule. Moreover, it has all the information required to rebuild a SYBYL molecule. Contrary to other fixed format files, Tripos Mol2 file is made in a free format that can be easily converted to another file format. Conventionally, MOL2 files describe the atom types, three dimensional coordinates of each atom, partial charges, types of bonds etc. The MOL2 format is frequently used to represent the three-dimensional structures of the ligands. Many molecular docking software requires ligand structure file in MOL2 format.

8.2 SDF (structured data file)

The structured data file (SDF) is a type of chemical-data file format that was designed by Biovia formerly known as Molecular Design Limited (MDL). The SDF provides two-dimensional or three-dimensional structural information of the ligands in plain text. Contrary to the MOL file which usually contains the structural information of one ligand, the SDF file format can hold single or multiple ligands separated by the sign of four dollars ($$$$). Additionally, SDF files also encode data on connectivity and hybridization state. SDF files are commonly used to represent ligand structures, especially in virtual screening studies.

8.3 PDBQT

The PDBQT (Protein Data Bank, Partial Charge, and Atom Type) format is considered an extension of the PDB (Protein Data Bank) file format. Overall, the PDBQT format is like the PDB format; however, it contains a bit more information for each of the ligand's atoms especially regarding their partial charges (q) and atom types (t).

The calculation of partial charges (q) is considered central for calculating the electrostatic interactions between ligands and receptors during the docking process carried out by the AutoDock tool. However, Autodock Vina does not rely on these charges for computing binding affinities. Rather, Vina considers the atom type (t) as a crucial variable in estimating the binding energies. It evaluates the role of atom types in ligand-receptor interactions as atom types are important in describing their behavior in the molecular docking process including interaction potentials and steric hindrances. PDBQT format can also be used to depict flexible compounds as it can also incorporate information about rotatable bonds so that the docking software can investigate the conformational space of the ligand during the docking process. In AutoDock-based molecular docking studies, PDBQT format is required for both the ligand and the receptor files.

8.4 XYZ (Cartesian coordinates)

The XYZ file format is known for its simplicity as it represents the total number of a ligand's atoms in the first row and a commentary in the second row. From the third row, different atoms are enlisted with their three-dimensional or Cartesian (X, Y, and Z) coordinates. XYZ files can be easily generated to be used in some docking software, especially for small ligands [54].

9. Active site prediction

The active site or binding site of an enzyme refers to the location where the substrate is fixed, which is typically situated on the surface of the enzyme. The term "ligand-binding region" refers to a segment of a protein or DNA/RNA molecule that is responsible for forming bonds with chemicals, drugs, or ions. This interaction results in a conformational change, ultimately leading to physiological, agonist, or antagonist effects [55]. The ligand-binding region is often characterized by a three-dimensional cleft or cavity that is shaped by the arrangement of amino acids in sequence.

According to recent research, it has been discovered that the central segments of numerous proteins have been nearly perfected for stability through natural selection. On the other hand, surface residues have not been optimized to the same extent, likely because protein function is facilitated through surface connections with other molecules. The proliferation of structural genomics initiatives has resulted in a substantial rise in the number of accessible protein structures. However, a significant portion of these structures are designated as hypothetical proteins due to the absence of biochemical information.

The experimental characterization of protein function is a labor-intensive and time-consuming process. To address this challenge, a computational tool for predicting functional sites in proteins would be invaluable. The value of such a tool is further emphasized by the automation required for structural genomics projects [56].

Based on the sequence or structural similarity of the query protein with well-characterized proteins, a variety of theoretical techniques are available to try and predict the activities of proteins. Proteins with identical sequences or structures may not always carry out comparable biological tasks. Similar tasks are also known to be carried out by proteins with distinct folds, such as subtilisin- and trypsin-like proteases [57]. Because of this disparity, structure-based methods have been developed, in which the similarity of the spatial arrangement of functionally important residues is used to predict function.

As a result, there is now more focus on determining the active site because it will impact the ligand's function and reaction. Energy functions, solvation models, and sampling algorithms capable of high-resolution structure prediction are necessary because binding and active sites are solvated or contain several polar and charged residues that create directed contacts to their ligands [58].

Following protein production, it is necessary to anticipate the protein's active site. Even though the receptor may have many active sites, only one of them needs to be selected. If heteroatoms and water molecules are present, they are mostly eliminated [59]. For carrying out molecular docking, we must be aware of the active site, either from literature research or active site prediction. We have no choice but to use the active site prediction if the literature does not provide sufficient evidence. Numerous online web servers are available to assist in predicting the active site, and some of them are listed below in **Table 2**.

S.No	Servers	links
1	CASTp	http://sts.bioengr.uic.edu/castp/
2	LigASite	http://www.bigre.ulb.ac.be/Users/benoit/LigASite/index.php?home
3	PDBeMotif	http://www.ebi.ac.uk/pdbe-site/pdbemotif/
4	*fPOP*	http://pocket.uchicago.edu/fpop/
5	metaPocket	http://metapocket.eml.org/
6	PocketQuer	http://pocketquery.csb.pitt.edu/
7	IBIS	http://www.ncbi.nlm.nih.gov/Structure/ibis/ibis.cgi
8	3DLigandStie	http://www.sbg.bio.ic.ac.uk/~3dligandsite/
9	SitesBase	http://www.modelling.leeds.ac.uk/sb/
10	PRO SURFER	http://163.42.136.93/top
11	KBDOCK	http://kbdock.loria.fr/index.php
12	Pocketome	http://www.pocketome.org/
13	sc-PDB	http://cheminfo.u-strasbg.fr:8080/scPDB/2011/db_search/about_scpdb.htm
14	ProBiS	http://probis.cmm.ki.si/index.php
15	Active Site prediction	http://www.scfbio-iitd.res.in/dock/ActiveSite_new.jsp
16	INDSITE	http://cssb.biology.gatech.edu/findsite
17	PocketDepth:	http://proline.physics.iisc.ernet.in/pocketdepth/
18	GHECOM 1.0	http://strcomp.protein.osaka-u.ac.jp/ghecom/
19	Pocket-Finder	http://www.modelling.leeds.ac.uk/pocketfinder/
20	Screen2	http://luna.bioc.columbia.edu/honiglab/screen2/cgi-bin/screen2.cgi
21	ConCavity	http://compbio.cs.princeton.edu/concavity/
22	ligplot	https://www.ebi.ac.uk/thornton-srv/software/LigPlus/

Table 2.
List of online protein active site predictors servers.

10. Recent advances and applications of molecular docking

Essentially, the goal of molecular docking is to use computational techniques to anticipate the structure of the ligand-receptor complex. The scientific community views this approach as a promising tool for drug discovery, especially for ligand identification, structure-based rational drug design, and prospective target predictions for many diseases. The remarkable advantages of molecular docking, specifically its capacity to predict experiments, are garnering increasing attention because of its possible uses in numerous industries [60].

10.1 In potential marine drugs discovery

Marine medications have been utilized for a long time and offer special benefits in clinical settings. Approximately 70% of the Earth's surface is made up of oceans, where a wide variety of marine species with high activity and efficacy levels have distinctive and innovative components not found on land. Research on marine pharmaceuticals has been its specialty since 1968 when the National Cancer Institute (NCI) began to examine marine resources for potential anti-cancer activity [61]. Marine drugs developed on this basis are important in the fields of antinociceptive, anti-bacterial, cerebrovascular diseases, neurological disorders, anti-virus, malaria, anti-coagulation, cardiovascular, and anti-tumor. The abundant marine organisms can produce a variety of naturally active substances with novel structures and remarkable activities [62].

Many marine-derived active compounds have been found and investigated globally in recent decades; many of them have either been licensed for commercialization or have progressed to different phases of clinical study. In addition to having high biological activity and a variety of unique structural features, marine species' metabolites also offer a wealth of pro-drugs and model structures for the study and creation of new medications. In recent decades, molecular docking technology a prominent technique in computer-aided drug design has been extensively utilized in the search for active ingredients and the clarification of action processes. It has also been crucial in the advancement of marine drug research and development. The in silico studies suggest that the halogenated compounds could exhibit anti-tyrosinase activity as their halogen group interacts with the catalytic hydrogen of the tyrosinase residues.

10.2 In food science

Several fundamental research about biomolecular interactions in the food matrix have gradually come to light in recent years. Since food gives the body vital nutrients including proteins, fats, carbs, and vitamins, molecular docking has a lot of applications in these fields. Furthermore, drug residues, biotoxins, and foodborne pathogens are major concerns related to food safety in food research, which is becoming more and more concerned with molecule-level analysis [16, 63–65]. The field of molecular docking has also marked its importance in discovering and identifying the natural compounds exhibiting the potential therapeutic effects or responsible for the toxic/side effects. Wang et al.'s work investigates how the binding of food items to enzymes can alter the structure and conformation of the enzyme, thereby lowering its activity [66]. Additionally, many food items compete with enzymes for substrate binding, which might lower the activity of the enzymes [67].

The docking studies investigate the antagonist/agonist or synergistic (positive or negative) reactions in the food-enzyme interaction study. The binding of lignin to α-amylase modifies the lignin structure, and the hydrogen bond identifies the hydrophobic cavity of the enzyme as the key binding site, explaining the activation of alpha-amylase. The primary binding force is the hydrophobic interaction, and the enzyme oversees the structural alterations. Understanding the relationship between an enzyme and its substrate as well as how enzyme activity is regulated is made easier by this [68]. Computational advancement also helped in exploring the bioactive peptides (BAPS) [69–71], amino acids [72–74], carbohydrates [75–78], lipids [79–81], vitamins [82–85], pesticides and veterinary drug residues [86–88], biotoxins [89–93], and foodborne pathogens [94–97].

10.3 In drug repurposing approach

Drug repurposing, or the use of already-approved medications for new uses, is a useful strategy for developing novel treatments for uncommon diseases while also cutting down on the time and expense associated with drug research. Identification of possible targets is an essential first step toward the search for novel indications. Molecule docking is a popular technique for locating possible targets. It may find possible targets for a given medicine or identify potential drugs for a specific target, and it only needs structure inputs from the drug and the target.

Molecular docking is widely used in drug development and repurposing, yet there are still issues with the technique. Adding co-binders to the system, taking solvents into consideration, and optimizing the target conformation are all potential ways to increase prediction accuracy.

The study conducted by Kumar et al. in 2017 applied this approach to identify the leading antipsychotic medications bromperidol, pimozide, anisoperidone, melperone, benperidol, and anisopirol as potential candidate for drug repurposing against a variety of AD-related targets and found that the most effective potential medication for interacting with many target proteins implicated in Alzhimer's disease is benperidol [98].

The COVID-19 pandemic has continued to bring consternation to most of the world, and the drug repurposing approach has found its application in identifying the drug against the virus. Many scientists applied the in silico approach to identify potential drugs by carrying out molecular docking on both approved drugs and substances previously tested *in vivo* to meet the time strain [99–101]. They have identified the molecules based on docking scores and found that the zafirlukast and simeprevir as the most promising candidates which were reported to exhibit leukotriene receptor antagonist and protease inhibitor [102].

Work done by Fortunatus et al., against schistosomiasis which is a neglected, yet common, disease of sub-tropical and tropical regions reveals that tolmetin and diflunisal are the potential drugs to be used against the disease even against the resistant species where the drugs like praziquantel or oxamniquine get failed [103]. Another study used this approach to find the best target against the Respiratory Syncytial Viral Infection, one of the most common viral infections of the respiratory airways having high mortality of younger children's and infants. The study suggested the use of oral hypoglycemic drugs (Glipizide, Glimepiride, Glibenclamide Canagloflozin,) as potential drug candidate for the condition requiring further validation by experimental, preclinical trials, clinical trials, and route of administration optimization [104].

Mile stonework was done by Baby K. et al. in 2023, against non-small cell lung carcinomas (NSCLC), a predominant form of lung malignancy and one of the highest morbidity rates due to the deregulation of Akt, a serine/threonine kinase. A computational investigation was undertaken to identify allosteric Akt-1 inhibitors from the previously FDA-approved drugs from the databases and found that the Akt-1 allosteric site showed highest binding affinity with the dasatinib, valganciclovir, novobiocin, and indacaterol [105]. Other examples of repurposed approved drugs are shown in **Table 3**.

Developed in the 1980s, itraconazole is a triazole antifungal medication that works well against a range of systemic fungal infections. It is recognized to hinder fungal development by impairing membrane activities and inhibiting the cytochrome P450-dependent lanosterol 14-α-demethylation (14DM) in the ergosterol biosynthesis pathway. Itraconazole's anticancer activity was initially documented by Chong et al. in 2007 because of its recently found anti-angiogenic activity.

Itraconazole had potent anticancer effects in preclinical models of medulloblastoma, basal cell carcinoma, and non-small cell lung cancer (NSCLC), either by alone or in conjunction with other anticancer medications. Itraconazole has entered multiple Phase II clinical investigations for the treatment of different forms of cancer, spurred by positive preclinical results. Positive clinical outcomes from trials for advanced lung cancer, prostate cancer, and basal cell carcinoma at Stanford University and Johns Hopkins Sidney Kimmel Comprehensive Cancer Center have been reported most recently. When compared to the control arm with pemetrexed

Drug	Original indication	Approved repurposed use
Allopurinol	Cancer	Gout
Aspirin	Inflammation, Pain	Antiplatelet
Amphotericin B	Antifungal	Antiparasitic
Baracitinib	Rheumatoid arthritis	COVID-19
Bromocriptine	Parkinson's disease	Diabetes mellitus
Bupropion	Depression	Smoking cessation
Duloxetine	Depression	Stress urinary incontinence
Doxycycline	Antibiotic	Antimalarial
Eflornithine	Antitumour	Antiparasitic
Finasteride	Benign prostatic hyperplasia	Hair loss
Gabapentin	Epilepsy	Neuropathic pain
Gemcitabine	Antiviral	Cancer
Infliximab	Crohn's Disease	Rheumatoid arthritis, Ulcerative colitis
Itraconazole	Antifungal	Anticancer
Methotrexate	Cancer	Rheumatoid arthritis
Miltefosine	Antitumour	Antiparasitic
Molnupiravir	Influenza viruses and encephalitic alphaviruses	COVID-19
Mycophenolate Mofetil	Prevention of organ transplant rejection	Lupus nephritis
Paromomycin	Antibiotic	Antiparasitic
Propranolol	Hypertension	Migraine headache
Raloxifene	Osteoporosis	Breast cancer
Remdesivir	Hepatitis–C	Ebola virus, COVID-19
Sildenafil	Angina	Erectile dysfunction
Thalidomide	Sedation, Morning sickness	Leprosy, Multiple myeloma
Tocilizumab	Rheumatoid arthritis, other autoimmune rheumatic diseases	COVID-19
Azidothymidine (Zidovudine)	Cancer	AIDS

Table 3.
Some examples of repurposed drugs [4, 106–116].

alone, patients with progressing nonsquamous non-small-cell lung cancer who had it in conjunction with itraconazole demonstrated a significant increase in survival.

Itraconazole has demonstrated encouraging anticancer potential against several cancer types; however, its exact anticancer mechanism is yet unknown. Itraconazole has been shown to have two anticancer mechanisms thus far: it inhibits the Hedgehog signaling system in some cancer cells and angiogenesis in certain cancer cells. To better understand the specific mechanism of action of itraconazole's anticancer effect and to aid in the development of a novel anticancer and anti-angiogenic medication, ongoing investigations are aimed at finding the molecular target of itraconazole in mammalian cells [111].

10.4 In overcoming scoring functions shortcomings through machine-learning

In silico drug discovery nowadays would not be possible without scoring functions. Still, it is a difficult challenge to accurately forecast binding affinity by scoring functions (SFs). The way scoring functions perform varies greatly depending on the target class. There is a great need for scoring systems that are based on accurate physics-based descriptors that better depict the protein-ligand recognition process. The predictive performance of classical structure-based virtual screening methodologies was either too poor or too varied, which led many practitioners to doubt its usefulness as a primary screening methodology for small compounds [117].

While widely used, classical SFs' performance is limited because they do not capture complex non-linear relationships or take advantage of the growing amount of structural data available to improve performance [118]. To address these issues, a number of machine learning-based scoring functions (MLSFs) have been developed over the last 10 years, utilizing convolutional neural networks [119], support vector machines [120], and random forests [121]. Additionally, while classical SFs are generally linear functions with empirical, force field-based, or knowledge-based terms to estimate binding affinity [122].

Though during the past 10 years, SFs' performance for virtual screening has improved, new approaches are required to push the envelope and make computer-driven drug discovery broadly usable. With so many experimental protein-ligand complex structures available, machine-learning scoring functions (MLSFs) appeared to have peaked in performance. Based on virtual screening benchmarks, MLSFs have shown significant performance gains over classical SFs by the prospective studies [123]. The majority of MLSFs are regressors, forecasting the receptor-ligand interaction's numerical pKd. To lower false positive rates during virtual screening, a classification approach to affinity prediction is also a feasible option [124].

Using the Support Vector Machine for Regression (SMOReg) and Random Forest (RF) algorithms, Guedes et al. developed nonlinear scoring functions that were trained using the same physics-based descriptors chosen for the final linear scoring functions [125]. on the proteases and protein-protein interactions (PPIs). They reported the first and well-performing SVM-based scoring function specific for PPI binding sites that can be a useful tool for finding new iPPIs by enhancing the solvation and ligand torsional entropy terms, implementing in MLR DockTScore predictions, DockTScore (available at www.dockthor.lncc.br) [122].

10.5 In GPU-system accelerated molecular docking

Computational drug development is greatly aided by virtual screening, which calls for precise and effective structure-based molecular docking. Many docking techniques have been developed and tested over the past 20 years [10, 15, 17, 37], but the most of them are primarily intended for CPU-based systems; just a small number [112, 114–115] are targeted at GPUs and heterogeneous HPC nodes [126, 127].

Fan et al. created molecular docking software building pieces and algorithms that utilize graphics processing units (GPUs) concentrating on MedusaDock [128], a versatile platform and method for protein-small molecule docking [129].

In addition to molecular docking-based HTVS, artificial intelligence (AI) methods are employed to investigate target and ligand binding potential. For instance, the MolAICal [130] program builds ligand molecules using a fragment growth technique, trains ligand fragment libraries for particular receptors using generative adversarial networks, and works with molecular docking software to determine the ideal conformation and binding energy.

Using a deep convolutional neural network technique, the DEEPScreen [131, 132]. software was trained individually for 704 targets using the SMILES of 2D molecules. Training involved using every receptor and at least 100 active ligands from the ChEMBL [133] chemical library. This resulted in the creation of unique predictive models for every target. Molecular docking-based HTVS, as opposed to AI approaches, continues to be the predominant choice in drug design due to the poor interpretability of AI methods and the limited accuracy of prediction outputs for molecules with major structural variations from the training molecules [134].

10.6 In large scale cloud docking

Molecular docking is a practical method to use structures to find ligands, and high throughput virtual (HTVS) screening is one of the basic parameters required for the approach in terms of finding a novel molecule. Large libraries are screened using this method, which quickly scores molecules for fit and ranks the database from best to worst for testing and purchase. The quantity of molecules that are commercially available has increased now by more than three orders of magnitude since 2015, opening up new possibilities for the discovery of novel chemistry and biology. After being used on more than a dozen targets, ultra-large-scale docking (LSD) has now found new compounds with activities that are frequently in the nano-molar and sporadically sub-nanomolar ranges. This has frequently resulted in molecules with intriguing in vivo activities [135].

Recent studies identifies that the docking computation cost for every target is embarrassingly high including the cost of licensed software, high GPU system and graphic system, meaning it may be distributed over several processors without significantly reducing performance [129].

Therefore, having a user-friendly and convenient online platform is crucial for regular users, particularly those who are unfamiliar with Linux operations, command lines, and other computer-based knowledge. Moreover, web services for molecular docking-based HTVS are a useful means of lowering the obstacles to HPC software utilization and accessibility while increasing productivity [136]. In addition to achieving notable gains in computational performance in recent years, public cloud computing platforms like Microsoft Azure, Google Cloud Platform, and Amazon Web Services are also anticipated to be capable of performing massively parallel computing. Bioinformatics is starting to use the cloud as it allows users to employ thousands of CPU cores and GPU accelerators arbitrarily, and because cloud images make it very easy to use different software kinds. As an example of an HPC cloud environment, Ohue et al. transferred the original protein-protein interaction prediction (protein–protein docking) software, MEGADOCK, onto Microsoft Azure and found it more efficient and cost effective [137].

Zhixiong et al. also indulged the cloud computing in marking the use of free energy perturbation (FEP) for binding affinity prediction between candidate compounds and their biological targets which has grown significantly. FEP applications do have certain drawbacks, though, such as their high cost, lengthy wait times, restricted scalability, and narrow range of application situations. Using enhanced simulation protocols, they have created XFEP, a scalable cloud computing platform for relative and absolute free energy forecasts, to address these issues. With the help of XFEP, large-scale FEP computations may be carried out more effectively, scalable, and affordably. For instance, 5000 compounds can be evaluated in a week utilizing 50–100 GPUs, with a computing cost that is nearly equal to the cost of synthesizing just one novel chemical.

Further, these abilities can be combined with artificial intelligence methods for goal-directed molecule generation and assessment to explore new avenues for applications in hit identification, hit-to-lead, and lead optimization drug discovery stages. These stages involve not only structure exploitation within the given chemical series but also evaluation and comparison of entirely unrelated molecules during structure exploration in a larger chemical space [138].

11. Conclusion

Making helpful predictions utilizing various tools that could be obtained through molecular modeling using computational techniques is one of the primary goals of computational medicinal chemistry. This fact is crucial when putting strategies into practice for drug development projects or protocols because of the various jobs and difficulties involved in finding the right hit compound. In this field, molecular docking plays a significant role and consistently offers benefits. Advances in computational power combined with imaging methods, artificial intelligence strategies, and protein fold prediction tools (e.g., AlphaFold), suggest a golden age for the application of the computational approaches to drug discovery, wherein knowledge of the function and shape of proteins within cells provides better protein structures for molecular docking, contributing to the development of targeted and selective new therapeutics while also lowering the expenses of testing.

Comparing repurposing molecules to traditional drug research and development approaches reveals certain advantages like Hasten's drug development, reduction in costs and risks, and personalized medicine while it still lacks the identification of the correct dose, potential pharmaceutical and toxicological issues, and the patenting rights. The importance of achieving notable gains in computational performance in recent years, through application of machine learning, by GPU system, overcoming the scoring function limitations, and public cloud computing platforms like Microsoft Azure, Google Cloud Platform, and Amazon Web Services in running large-scale molecular docking screens is now pragmatic and does not require any commercial software exploring new avenues for applications in hit identification, hit-to-lead, and lead optimization drug discovery stages.

Conflict of interest

"The authors declare no conflict of interest."

Glossary

FDA	Food and Drug Administration
COVID-19	coronavirus disease 2019
HTS	high-throughput screening
NCBI	National Center for Biotechnology Information
GA	genetic algorithm
ADAM	advanced design and modeling
FRED	fast rigid exhaustive docking
GLIDE	grid-based ligand docking from energetics

SLIDE	screening for ligands by induced-fit docking, efficiently
GOLD	genetic optimization for ligand docking
MOE	molecular operating environment
PLANTS	protein-ligand ANT system
QM	quantum mechanics
SDF	structured data file
PDBQT	protein data bank, partial charge, and atom type partial charges (q) and atom types (t)
XYZ Coordinates	Cartesian coordinates
DNA	deoxyribonucleic acid
RNA	ribonucleic acid
BAPS	bioactive peptides
AD	Alzheimer's disease
NSCLC	non-small cell lung carcinomas
	Drug repurposing or drug repositioning or drug reprofiling the technique of using an existing approved drug candidate for a new treatment or medical indication for which it was not indicated before
Serendipity	accidental, unexpected, and fortunate discoveries
Off-label use	drug used for different purposes than those for which they were initially licensed
Molecular docking	interaction of two or more molecules to give the stable adduct
De Novo drug development	methodology that creates novel chemical entities based only on the information regarding a biological target (receptor) or its known active binders
Drug design	the approach of finding drugs by design, based on their biological targets
Drug discovery	process of identifying chemical entities that have the potential to become therapeutic agents

Author details

Sana Shamim[1]*, Rabya Munawar[1], Yasmeen Rashid[2], Sheikh Muhammad Zesshan Qadar[3], Rabia Bushra[4], Irshad Begum[5], Muhammad Imran[6] and Tehseen Quds[7]

1 Dow College of Pharmacy, Faculty of Pharmaceutical Sciences, Dow University of Health Sciences, Ojha Campus, Department of Pharmaceutical Chemistry, Karachi, Pakistan

2 Department of Biochemistry, University of Karachi, Karachi, Pakistan

3 The National Collaborating Centre for Infectious Diseases, University of Manitoba, Winnipeg, MB, Canada

4 Dow College of Pharmacy, Faculty of Pharmaceutical Sciences, Dow University of Health Sciences, Ojha Campus, Department of Pharmaceutics, Karachi, Pakistan

5 Department of Chemistry, University of Karachi, Karachi, Pakistan

6 Department of Pharmacy, Iqra University, Karachi, Pakistan

7 Department of Pharmacognosy, Dow College of Pharmacy, Faculty of Pharmaceutical Sciences, Dow University of Health Sciences, Ojha Campus, Karachi, Pakistan

*Address all correspondence to: sana.shamim@duhs.edu.pk

IntechOpen

References

[1] DiMasi JA, Grabowski HG, Hansen RW. Innovation in the pharmaceutical industry: New estimates of R&D costs. Journal of Health Economics. 2016;**47**:20-33. DOI: 10.1016/j.jhealeco.2016.01.012

[2] Law GL, Tisoncik-Go J, Korth MJ, Katze MG. Drug repurposing: A better approach for infectious disease drug discovery? Current Opinion in Immunology. 2013;**25**(5):588-592. DOI: 10.1016/j.coi.2013.08.004

[3] Kumar S, Roy V. Repurposing drugs: An empowering approach to drug discovery and development. Drug Research. 2023;**73**(9):481-490. DOI: 10.1055/a-2095-0826

[4] Park K. A review of computational drug repurposing. Translational and Clinical Pharmacology. 2019;**27**(2):59-63. DOI: 10.12793/tcp.2019.27.2.59

[5] Oprea TI, Bauman JE, Bologa CG, Buranda T, Chigaev A, Edwards BS, et al. Drug repurposing from an academic perspective. Drug Discovery Today: Therapeutic Strategies. 2011;**8**(3-4):61-69. DOI: 10.1016/j.ddstr.2011.10.002

[6] Ding H, Takigawa I, Mamitsuka H, Zhu S. Similarity-based machine learning methods for predicting drug–target interactions: A brief review. Briefings in Bioinformatics. 2014;**15**(5):734-747. DOI: 10.1093/bib/bbt056

[7] Cha Y, Erez T, Reynolds IJ, Kumar D, Ross J, Koytiger G, et al. Drug repurposing from the perspective of pharmaceutical companies. British Journal of Pharmacology. 2018;**175**(2):168-180. DOI: 10.1111/bph.13798

[8] Fan J, Fu A, Zhang L. Progress in molecular docking. Quantitative Biology. 2019;**7**:83-89. DOI: 10.1007/s40484-019-0172-y

[9] Wu G, Robertson DH, Brooks CL III, Vieth M. Detailed analysis of grid-based molecular docking: A case study of CDOCKER—A CHARMm-based MD docking algorithm. Journal of Computational Chemistry. 2003;**24**(13):1549-1562. DOI: 10.1002/jcc.10306

[10] Emig D, Ivliev A, Pustovalova O, Lancashire L, Bureeva S, Nikolsky Y, et al. Drug target prediction and repositioning using an integrated network-based approach. PLoS One. 2013;**8**(4):e60618. DOI: 10.1371/journal.pone.0060618

[11] Swamidass SJ. Mining small-molecule screens to repurpose drugs. Briefings in Bioinformatics. 2011;**12**(4):327-335. DOI: 10.1093/bib/bbr028

[12] Doman TN, McGovern SL, Witherbee BJ, Kasten TP, Kurumbail R, Stallings WC, et al. Molecular docking and high-throughput screening for novel inhibitors of protein tyrosine phosphatase-1B. Journal of Medicinal Chemistry. 2002;**45**(11):2213-2221. DOI: 10.1021/jm010548w

[13] Jin G, Fu C, Zhao H, Cui K, Chang J, Wong ST. A novel method of transcriptional response analysis to facilitate drug repositioning for cancer therapy. Cancer Research. 2012;**72**(1):33-44. DOI: 10.1158/0008-5472

[14] Haeberle H, Dudley JT, Liu JT, Butte AJ, Contag CH. Identification of cell surface targets through meta-analysis of microarray data. Neoplasia. 2012;**14**(7):666-669. DOI: 10.1593/neo.12634

[15] Hebbring SJ. The challenges, advantages and future of phenome-wide association studies. Immunology. 2014;**141**:157-165. DOI: 10.1111/imm.12195

[16] Tao X, Huang Y, Wang C, Chen F, Yang L, Ling L, et al. Recent developments in molecular docking technology applied in food science: A review. International Journal of Food Science & Technology. 2020;**55**(1):33-45. DOI: 10.1111/ijfs.14325

[17] Ferreira LG, Dos Santos RN, Oliva G, Andricopulo AD. Molecular docking and structure-based drug design strategies. Molecules. 2015;**20**(7):13384-13421. DOI: 10.3390/molecules200713384

[18] Forli S, Huey R, Pique ME, Sanner MF, Goodsell DS, Olson AJ. Computational protein–ligand docking and virtual drug screening with the AutoDock suite. Nature Protocols. 2016;**11**(5):905-919. DOI: 10.1038/nprot.2016.051

[19] Roche DB, Brackenridge DA, McGuffin LJ. Proteins and their interacting partners: An introduction to protein–ligand binding site prediction methods. International Journal of Molecular Sciences. 2015;**16**(12):29829-29842. DOI: 10.3390/ijms161226202

[20] Jain AN, Nicholls A. Recommendations for evaluation of computational methods. Journal of Computer-aided Molecular Design. 2008;**22**:133-139. DOI: 10.1007/s10822-008-9196-5

[21] Guedes IA, de Magalhães CS, Dardenne LE. Receptor–ligand molecular docking. Biophysical Reviews. 2014;**6**:75-87. DOI: 10.1007/s12551-013-0130-2

[22] Elokely KM, Doerksen RJ. Docking challenge: Protein sampling and molecular docking performance. Journal of Chemical Information and Modeling. 2013;**53**(8):1934-1945. DOI: 10.1021/ci400040d

[23] Yuriev E, Ramsland PA. Latest developments in molecular docking: 2010-2011 in review. Journal of Molecular Recognition. 2013;**26**(5):215-239. DOI: 10.1002/jmr.2266

[24] Cross JB, Thompson DC, Rai BK, Baber JC, Fan KY, Hu Y, et al. Comparison of several molecular docking programs: Pose prediction and virtual screening accuracy. Journal of Chemical Information and Modeling. 2009;**49**(6):1455-1474. DOI: 10.1021/ci900056c

[25] Thomsen R, Christensen MH. MolDock: A new technique for high-accuracy molecular docking. Journal of Medicinal Chemistry. 2006;**49**(11):3315-3321. DOI: 10.1021/jm051197e

[26] Raval K, Ganatra T. Basics, types and applications of molecular docking: A review. IP International Journal of Comprehensive and Advanced Pharmacology. 2022;**7**(1):12-16. DOI: 10.18231/j.ijcaap.2022.003

[27] Morrison JL, Breitling R, Higham DJ, Gilbert DR. A lock-and-key model for protein–protein interactions. Bioinformatics. 2006;**22**(16):2012-2019. DOI: 10.1093/bioinformatics/btl338

[28] Koshland DE Jr. The key–lock theory and the induced fit theory. Angewandte Chemie International Edition in English. 1995;**33**(23-24):2375-2378. DOI: 10.1002/anie.199423751

[29] Mizuguchi T, Matubayasi N. Free-energy analysis of peptide binding in lipid membrane using all-atom molecular dynamics simulation combined with theory of solutions. The Journal of Physical Chemistry B. 2018;**122**(13):3219-3229. DOI: 10.1021/acs.jpcb.7b08241

[30] Zhang N, Huo J, Yang B, Ruan X, Zhang X, Bao J, et al. Understanding of imidazolium group hydration and polymer structure for hydroxide anion conduction in hydrated imidazolium-g-PPO membrane by molecular dynamics simulations. Chemical Engineering Science. 2018;**192**:1167-1176. DOI: 10.1016/j.ces.2018.08.051

[31] Arthur DE, Uzairu A, Mamza P, Abechi SE, Shallangwa GA. Structure-based optimization of tyrosine kinase inhibitors: A molecular docking study. Network Modeling Analysis in Health Informatics and Bioinformatics. 2018;**7**:1-8. DOI: 10.1007/s13721-018-0170-4

[32] Nascimento KS, Araripe DA, Pinto-Junior VR, Osterne VJ, Martins FW, Neco AH, et al. Homology modeling, molecular docking, and dynamics of two α-methyl-d-mannoside-specific lectins from Arachis genus. Journal of Molecular Modeling. 2018;**24**:1-10. DOI: 10.1007/s00894-018-3800-y

[33] Nie X, Zhao L, Deng S, Su W, Zhang Y. A review of molecular simulation applied in vapor-liquid equilibria (VLE) estimation of thermodynamic cycles. Journal of Molecular Liquids. 2018;**264**:652-674. DOI: 10.1016/j.molliq.2018.05.101

[34] Shoichet BK, McGovern SL, Wei B, Irwin JJ. Lead discovery using molecular docking. Current Opinion in Chemical Biology. 2002;**6**(4):439-446. DOI: 10.1016/S1367-5931(02)00339-3

[35] Gschwend DA, Good AC, Kuntz ID. Molecular docking towards drug discovery. Journal of Molecular Recognition: An Interdisciplinary Journal. 1996;**9**(2):175-186. DOI: 10.1002/(SICI)1099-1352(199603)9:2<175::AID-JMR260>3.0.CO;2-D

[36] Agarwal S, Mehrotra RJ. An overview of molecular docking. JSM Chemistry. 2016;**4**(2):1024-1028

[37] Yan Z, Wang J. Optimizing scoring function of protein-nucleic acid interactions with both affinity and specificity. PLoS One. 2013;**8**(9):e74443. DOI: 10.1371/journal.pone.0074443

[38] Liu W, Liu G, Zhou H, Fang X, Fang Y, Wu J. Computer prediction of paratope on antithrombotic antibody 10B12 and epitope on platelet glycoprotein VI via molecular dynamics simulation. BioMedical Engineering OnLine. 2016;**15**(2):647-658. DOI: 10.1186/s12938-016-0272-0

[39] Roy S, Narang BK, Gupta MK, Abbot V, Singh V, Rawal RK. Molecular docking studies on isocytosine analogues as xanthine oxidase inhibitors. Drug Research. 2018;**68**(07):395-402. DOI: 10.1055/s-0043-125210

[40] Tripathi A, Misra K. Molecular docking: A structure-based drug designing approach. JSM Chemistry. 2017;**5**(2):1042-1047

[41] Morris GM, Goodsell DS, Halliday RS, Huey R, Hart WE, Belew RK, et al. Automated docking using a Lamarckian genetic algorithm and an empirical binding free energy function. Journal of Computational Chemistry. 1998;**19**(14):1639-1662. DOI: 10.1002/(SICI)1096-987X(19981115)19:14<1639::AID-JCC10>3.0.CO;2-B

[42] Tiwari G, Mohanty D. An in silico analysis of the binding modes and binding affinities of small molecule modulators of PDZ-peptide interactions. PLoS One. 2013;**8**(8):e71340. DOI: 10.1371/journal.pone.0071340

[43] Lorber DM, Shoichet BK. Flexible ligand docking using conformational ensembles. Protein Science. 1998;**7**(4):938-950. DOI: 10.1002/pro.5560070411

[44] Huang SY, Zou X. Ensemble docking of multiple protein structures: Considering protein structural variations in molecular docking. Proteins: Structure, Function, and Bioinformatics. 2007;**66**(2):399-421. DOI: 10.1002/prot.21214

[45] Morris GM, Lim-Wilby M. Molecular docking. In: Kukol A, editor. Molecular Modeling of Proteins. Totowa, NJ: Humana Press; 2008. 365-382p

[46] Huang SY, Grinter SZ, Zou X. Scoring functions and their evaluation methods for protein-ligand docking: Recent advances and future directions. Physical Chemistry Chemical Physics. 2010;**12**(40):12899-12908. DOI: 10.1039/c0cp00151a

[47] Englebienne P, Moitessier N. Docking ligands into flexible and solvated macromolecules. 5. Force-field-based prediction of binding affinities of ligands to proteins. Journal of Chemical Information and Modeling. 2009;**49**(11):2564-2571. DOI: 10.1021/ci900251k

[48] Kitchen DB, Decornez H, Furr JR, Bajorath J. Docking and scoring in virtual screening for drug discovery: Methods and applications. Nature Reviews. Drug Discovery. 2004;**3**(11):935-949. DOI: 10.1038/nrd1549

[49] Jain AN. Surflex: Fully automatic flexible molecular docking using a molecular similarity-based search engine. Journal of Medicinal Chemistry. 2003;**46**(4):499-511. DOI: 10.1021/jm020406h

[50] Rarey M, Kramer B, Lengauer T, Klebe G. A fast flexible docking method using an incremental construction algorithm. Journal of Molecular Biology. 1996;**261**(3):470-489. DOI: 10.1006/jmbi.1996.0477

[51] Sotriffer C. Docking of covalent ligands: Challenges and approaches. Molecular Informatics. 2018;**37**(9-10):e1800062. DOI: 10.1002/minf.201800062

[52] Klebe G. Virtual ligand screening: Strategies, perspectives and limitations. Drug Discovery Today. 2006;**11**(13-14):580-594. DOI: 10.1016/j.drudis.2006.05.012

[53] Chen YC. Beware of docking! Trends in Pharmacological Sciences. 2015;**36**(2):78-95. DOI: 10.1016/j.tips.2014.12.001

[54] Stourac J, Dubrava J, Musil M, Horackova J, Damborsky J, Mazurenko S, Bednar D. FireProtDB: Database of manually curated protein stability data. Nucleic Acids Research. 2021;**49**(D1):D319-24. 10.1093/nar/gkaa981

[55] Fanelli A, Sullivan ML. Tools for protein structure prediction and for molecular docking applied to enzyme active site analysis: A case study using a BAHD hydroxycinnamoyl transferase In: Methods in Enzymology. Vol. 683. Cambridge, MA, US: Academic Press; 2023. pp. 41-79

[56] Yang YH, Ku X, Gong YN, Meng FL, Dongbo DP, Guo YH, et al. Prediction of superantigen active sites and clonal expression of staphylococcal enterotoxin-like W. Zhonghua liu Xing Bing xue za zhi= Zhonghua Liuxingbingxue Zazhi. 2023;**44**(4):629-635. DOI: 10.3760/cma.j.cn112338-20220822-00725

[57] Liao J, Wang Q, Wu F, Huang Z. In Silico methods for identification of potential active sites of therapeutic targets. Molecules. 2022;**27**(20):7103. DOI: 10.3390/molecules27207103

[58] Mercado J. Thermodynamic Stability of Loop 6 Motion in Human

Triosephosphate Isomerase Variants [doctoral dissertation] California State University, Long Beach; 2023

[59] Chaudhary KK, Mishra N. A review on molecular docking: Novel tool for drug discovery. Database. 2016;**3**(4):1029. ISSN: 2333-6633

[60] Zhao Y, Zeng H, Zhu XW, Lu W, Li D. Metal–organic frameworks as photoluminescent biosensing platforms: Mechanisms and applications. Chemical Society Reviews. 2021;**50**(7):4484-4513. DOI: 10.1039/D0CS00955E

[61] Chen G, Seukep AJ, Guo M. Recent advances in molecular docking for the research and discovery of potential marine drugs. Marine Drugs. 2020;**18**(11):545. DOI: 10.3390/md18110545

[62] Paudel P, Wagle A, Seong SH, Park HJ, Jung HA, Choi JS. A new tyrosinase inhibitor from the red alga Symphyocladia latiuscula (Harvey) Yamada (Rhodomelaceae). Marine Drugs. 2019;**17**(5):295. DOI: 10.3390/md17050295

[63] San-Martin A, Donoso V, Leiva S, Bacho M, Nunez S, Gutierrez M, et al. Molecular docking studies of the antitumoral activity and characterization of new chalcone. Current Topics in Medicinal Chemistry. 2015;**15**(17):1743-1749

[64] Nakano S, Megro SI, Hase T, Suzuki T, Isemura M, Nakamura Y, et al. Computational molecular docking and X-ray crystallographic studies of catechins in new drug design strategies. Molecules. 2018;**23**(8):2020. DOI: 10.3390/molecules23082020

[65] Śledź P, Caflisch A. Protein structure-based drug design: From docking to molecular dynamics. Current Opinion in Structural Biology. 2018;**48**:93-102. DOI: 10.1016/j.sbi.2017.10.010

[66] Wang J, Chan C, Huang FW, Xie JF, Xu H, Ho KW, et al. Interaction mechanism of pepsin with a natural inhibitor gastrodin studied by spectroscopic methods and molecular docking. Medicinal Chemistry Research. 2017;**26**:405-413. DOI: 10.1007/s00044-016-1760-2

[67] Zhao L, Guo R, Sun Q, Lan J, Li H. Interaction between azo dye acid red 14 and pepsin by multispectral methods and docking studies. Luminescence. 2017;**32**(7):1123-1130. DOI: 10.1002/bio.3298

[68] Xie F, Zhang W, Gong S, Gu X, Lan X, Wu J, et al. Investigating lignin from Canna edulis ker residues induced activation of α-amylase: Kinetics, interaction, and molecular docking. Food Chemistry. 2019;**271**:62-69. DOI: 10.1016/j.foodchem.2018.07.153

[69] Agrawal H, Joshi R, Gupta M. Purification, identification and characterization of two novel antioxidant peptides from finger millet (Eleusine coracana) protein hydrolysate. Food Research International. 2019;**120**:697-707. DOI: 10.1016/j.foodres.2018.11.028

[70] Xue Z, Wen H, Zhai L, Yu Y, Li Y, Yu W, et al. Antioxidant activity and anti-proliferative effect of a bioactive peptide from chickpea (Cicer arietinum L.). Food Research International. 2015;**77**:75-81. DOI: 10.1016/j.foodres.2015.09.027

[71] Dang Y, Hao L, Cao J, Sun Y, Zeng X, Wu Z, et al. Molecular docking and simulation of the synergistic effect between umami peptides, monosodium glutamate and taste receptor T1R1/T1R3. Food Chemistry. 2019;**271**:697-706. DOI: 10.1016/j.foodchem.2018.08.001

[72] Li J, Geng S, Liu B, Wang H, Liang G. Self-assembled mechanism of hydrophobic amino acids and β-cyclodextrin based on experimental and computational methods. Food Research International. 2018;**112**:136-142. DOI: 10.1016/j.foodres.2018.06.017

[73] Hartmann A, Gostner J, Fuchs J, Chaita E, Aligiannis N, Skaltsounis L, et al. Inhibition of collagenase by mycosporine-like amino acids from marine sources. Planta Medica. 2015;**81**(10):813-820. DOI: 10.1055/s-0035-1546105

[74] Dettori L, Jelsch C, Guiavarc'h Y, Delaunay S, Framboisier X, Chevalot I, et al. Molecular rules for selectivity in lipase-catalysed acylation of lysine. Process Biochemistry. 2018;**74**:50-60. DOI: 10.1016/j.procbio.2018.07.021

[75] Goel A, Gajula K, Gupta R, Rai B. In-silico prediction of sweetness using structure-activity relationship models. Food Chemistry. 2018;**253**:127-131. DOI: 10.1016/j.foodchem.2018.01.111

[76] Yang JP, He H, Lu YH. Four flavonoid compounds from Phyllostachys edulis leaf extract retard the digestion of starch and its working mechanisms. Journal of Agricultural and Food Chemistry. 2014;**62**(31):7760-7770. DOI: 10.1021/jf501931m

[77] Kato-Schwartz CG, Bracht F, de Almeida GG, Soares AA, Vieira TF, Brugnari T, et al. Inhibition of α-amylases by pentagalloyl glucose: Kinetics, molecular dynamics and consequences for starch absorption. Journal of Functional Foods. 2018;**44**:265-273. DOI: 10.1016/j.jff.2018.03.025

[78] Wang Y, Wang Y, Luo Q, Zhang H, Cao J. Molecular characterization of the effects of Ganoderma Lucidum polysaccharides on the structure and activity of bovine serum albumin. Spectrochimica Acta Part A: Molecular and Biomolecular Spectroscopy. 2019;**206**:538-546. DOI: 10.1016/j.saa.2018.08.051

[79] Rong C, Chen H, Tang X, Gu Z, Zhao J, Zhang H, et al. Characterization and molecular docking of new Δ17 fatty acid desaturase genes from Rhizophagus irregularis and Octopus bimaculoides. RSC Advances. 2019;**9**(12):6871-6880. DOI: 10.1039/C9RA00535H

[80] Zhu Z, Chen J, Wang G, Elsherbini A, Zhong L, Jiang X, et al. Ceramide regulates interaction of Hsd17b4 with Pex5 and function of peroxisomes. Biochimica et Biophysica Acta (BBA)-Molecular and Cell Biology of Lipids. 2019;**1864**(10):1514-1524. DOI: 10.1016/j.bbalip.2019.05.017

[81] El Shatshat A, Pham AT, Rao PP. Interactions of polyunsaturated fatty acids with amyloid peptides Aβ40 and Aβ42. Archives of Biochemistry and Biophysics. 2019;**663**:34-43. DOI: 10.1016/j.abb.2018.12.027

[82] de Oliveira VR, Domingueti CP. Association of vitamin D deficiency and type 1 diabetes mellitus: A systematic review and meta-analysis. International Journal of Diabetes in Developing Countries. 2018;**38**:280-288. DOI: 10.1007/s13410-018-0607-4

[83] Abdi F, Movahedi M, Nikje MA, Ghanei L, Mirzaie S. Vitamin D as a modulating agent of metformin and insulin in patients with type 2 diabetes. Journal of Research in Pharmacy. 2019;**23**:360-378. DOI: 10.12991/jrp.2019.144

[84] Lin S, Zhang G, Liao Y, Gong D. The inhibitory kinetics and mechanism of dietary vitamins D 3 and B 2 on xanthine oxidase. Food & Function.

2016;**7**(6):2849-2861. DOI: 10.1039/C6FO00491A

[85] Borah PK, Sarkar A, Duary RK. Water-soluble vitamins for controlling starch digestion: Conformational scrambling and inhibition mechanism of human pancreatic α-amylase by ascorbic acid and folic acid. Food Chemistry. 2019;**288**:395-404. DOI: 10.1016/j.foodchem.2019.03.022

[86] Lan J, Zhao H, Jin X, Guan H, Song Y, Fan Y, et al. Development of a monoclonal antibody-based immunoaffinity chromatography and a sensitive immunoassay for detection of spinosyn a in milk, fruits, and vegetables. Food Control. 2019;**95**:196-205. DOI: 10.1016/j.foodcont.2018.08.002

[87] Sok V, Fragoso A. Kinetic, spectroscopic and computational docking study of the inhibitory effect of the pesticides 2, 4, 5-T, 2, 4-D and glyphosate on the diphenolase activity of mushroom tyrosinase. International Journal of Biological Macromolecules. 2018;**118**:427-434. DOI: 10.1016/j.ijbiomac.2018.06.098

[88] Wang G, Zhang HC, Liu J, Wang JP. A receptor-based chemiluminescence enzyme linked immunosorbent assay for determination of tetracyclines in milk. Analytical Biochemistry. 2019;**564**:40-46. DOI: 10.1016/j.ab.2018.10.017

[89] Poór M, Kunsági-Máté S, Bálint M, Hetényi C, Gerner Z, Lemli B. Interaction of mycotoxin zearalenone with human serum albumin. Journal of Photochemistry and Photobiology B: Biology. 2017;**170**:16-24. DOI: 10.1016/j.jphotobiol.2017.03.016

[90] Poór M, Lemli B, Bálint M, Hetényi C, Sali N, Kőszegi T, et al. Interaction of citrinin with human serum albumin. Toxins. 2015;7(12):5155-5166. DOI: 10.3390/toxins7124871

[91] Argudín MÁ, Mendoza MC, Rodicio MR. Food poisoning and Staphylococcus aureus enterotoxins. Toxins. 2010;**2**(7):1751-1773. DOI: 10.3390/toxins2071751

[92] Wu D, Duan R, Geng F, Hu X, Gan N, Li H. Comparative analysis of the interaction of mono-, dis-, and tris-azo food dyes with egg white lysozyme: A combined spectroscopic and computational simulation approach. Food Chemistry. 2019;**284**:180-187. DOI: 10.1016/j.foodchem.2019.01.115

[93] Zulfakar MH, Chan LM, Rehman K, Wai LK, Heard CM. Coenzyme Q10-loaded fish oil-based bigel system: Probing the delivery across porcine skin and possible interaction with fish oil fatty acids. AAPS PharmSciTech. 2018:1116-1123. DOI: 10.1208/s12249-017-0923-x

[94] Agrawal S, Kulabhusan PK, Joshi M, Bodas D, Paknikar KM. A high affinity phage-displayed peptide as a recognition probe for the detection of salmonella Typhimurium. Journal of Biotechnology. 2016;**231**:40-45. DOI: 10.1016/j.jbiotec.2016.05.02

[95] Hossain MM, Roy PK, Mosnaz AT, Shakil SK, Hasan MM, Prodhan SH. Structural analysis and molecular docking of potential ligands with chorismate synthase of listeria monocytogenes: A novel antibacterial drug target. Indian Journal of Biochemistry & Biophysics. 2015;**52**:45-59

[96] Kar S, Mishra RK, Pathak A, Dikshit A, Golakoti NR. In silico modeling and synthesis of phenyl and thienyl analogs of chalcones for potential leads as anti-bacterial agents. Journal of Molecular Structure. 2018;**1156**:433-440. DOI: 10.1016/j.molstruc.2017.12.002

[97] Geethalakshmi R, Sarada VD. In vitro and in silico antimicrobial activity

of sterol and flavonoid isolated from Trianthema decandra L. Microbial Pathogenesis. 2018;**121**:77-86. DOI: 10.1016/j.micpath.2018.05.018

[98] Kumar S, Chowdhury S, Kumar S. In silico repurposing of antipsychotic drugs for Alzheimer's disease. BMC Neuroscience. 2017;**18**(1):1-6. DOI: 10.1186/s12868-017-0394-8

[99] Shah B, Modi P, Sagar SR. In silico studies on therapeutic agents for COVID-19: Drug repurposing approach. Life Sciences. 2020;**252**:117652. DOI: 10.1016/j.lfs.2020.117652

[100] Cavasotto CN, Di Filippo JI. In Silico drug repurposing for COVID-19: Targeting SARS-CoV-2 proteins through docking and consensus ranking. Molecular Informatics. 2021;**40**(1):2000115. DOI: 10.1002/minf.202000115

[101] Li Q, Kang C. Progress in developing inhibitors of SARS-CoV-2 3C-like protease. Microorganisms. 2020;**8**(8):1250. DOI: 10.3390/microorganisms8081250

[102] Lazniewski M, Dermawan D, Hidayat S, Muchtaridi M, Dawson WK, Plewczynski D. Drug repurposing for identification of potential spike inhibitors for SARS-CoV-2 using molecular docking and molecular dynamics simulations. Methods. 2022;**203**:498-510. DOI: 10.1016/j.ymeth.2022.02.004

[103] Ezebuo FC, Uzochukwu IC. Drug repurposing for schistosomiasis: Molecular docking and dynamics investigations. Journal of Biomolecular Structure and Dynamics. 2022;**40**(3):995-1009. DOI: 10.1080/07391102.2020.1820382

[104] Adediran EO. Repurposing antidiabetic drugs against respiratory syncytial viral infection: A docking study. Computational Molecular Bioscience. 2022;**12**(2):85-94. DOI: 10.4236/cmb.2022.122005

[105] Baby K, Maity S, Mehta CH, Nayak UY, Shenoy GG, Pai KS, et al. Computational drug repurposing of Akt-1 allosteric inhibitors for non-small cell lung cancer. Scientific Reports. 2023;**13**(1):7947. DOI: 10.1038/s41598-023-35122-7

[106] Meyerhoff A. US Food and Drug Administration approval of AmBisome (liposomal amphotericin B) for treatment of visceral leishmaniasis. Clinical Infectious Diseases. 1999;**28**(1):42-48. DOI: 10.1086/515085

[107] Rodrigues L, Bento Cunha R, Vassilevskaia T, Viveiros M, Cunha C. Drug repurposing for COVID-19: A review and a novel strategy to identify new targets and potential drug candidates. Molecules. 2022;**27**(9):2723. DOI: 10.3390/molecules27092723

[108] Tan KR, Magill AJ, Parise ME, Arguin PM. Doxycycline for malaria chemoprophylaxis and treatment: Report from the CDC expert meeting on malaria chemoprophylaxis. The American Journal of Tropical Medicine and Hygiene. 2011;**84**(4):517-531. DOI: 10.4269/ajtmh.2011.10-0285

[109] Simarro PP, Franco J, Diarra A, Postigo JR, Jannin J. Update on field use of the available drugs for the chemotherapy of human African trypanosomiasis. Parasitology. 2012;**139**(7):842-846. DOI: 10.1017/S0031182012000169

[110] Ten CF. Years of infliximab (Remicade®) in clinical practice: The story from bench to bedside. European Journal of Pharmacology. 2009;**623**:S1-S4. DOI: 10.1016/j.ejphar.2009.10.023

[111] Shim JS, Liu JO. Recent advances in drug repositioning for the discovery of new anticancer drugs. International Journal of Biological Sciences. 2014;**10**(7):654-663. DOI: 10.7150/ijbs.9224

[112] Smorenburg CH, Seynaeve C, Bontenbal M, Planting AS, Sindermann H, Verweij J. Phase II study of miltefosine 6% solution as topical treatment of skin metastases in breast cancer patients. Anti-Cancer Drugs. 2000;**11**(10):825-828

[113] Regueira TB, Kildegaard KR, Hansen BG, Mortensen UH, Hertweck C, Nielsen J. Molecular basis for mycophenolic acid biosynthesis in Penicillium brevicompactum. Applied and Environmental Microbiology. 2011;**77**(9):3035-3043. DOI: 10.1128/AEM.03015-10

[114] Ben Salah A, Ben Messaoud N, Guedri E, Zaatour A, Ben Alaya N, Bettaieb J, et al. Topical paromomycin with or without gentamicin for cutaneous leishmaniasis. New England Journal of Medicine. 2013;**368**(6):524-532. DOI: 10.1056/NEJMoa1202657

[115] Beigel JH, Tomashek KM, Dodd LE, Mehta AK, Zingman BS, Kalil AC, et al. Remdesivir for the treatment of Covid-19. New England Journal of Medicine. 2020;**383**(19):1813-1826. DOI: 10.1056/NEJMoa2007764

[116] Volberding PA, Lagakos SW, Koch MA, Pettinelli C, Myers MW, Booth DK, et al. Zidovudine in asymptomatic human immunodeficiency virus infection: A controlled trial in persons with fewer than 500 CD4-positive cells per cubic millimeter. New England Journal of Medicine. 1990;**322**(14):941-949. DOI: 10.1056/NEJM199004053221401

[117] Martin EJ, Sullivan DC. Surrogate AutoShim: Predocking into a universal ensemble kinase receptor for three dimensional activity prediction, very quickly, without a crystal structure. Journal of Chemical Information and Modeling. 2008;**48**(4):873-881. DOI: 10.1021/ci700455u

[118] Adeshina YO, Deeds EJ, Karanicolas J. Machine learning classification can reduce false positives in structure-based virtual screening. National Academy of Sciences of the United States of America. 2020;**117**(31):18477-18488. DOI: 10.1073/pnas.2000585117

[119] Pason LP, Sotriffer CA. Empirical scoring functions for affinity prediction of protein-ligand complexes. Molecular Informatics. 2016;**35**(11-12):541-548. DOI: 10.1002/minf.201600048

[120] Greener JG, Kandathil SM, Moffat L, Jones DT. A guide to machine learning for biologists. Nature Reviews Molecular Cell Biology. 2022;**23**(1):40-55. DOI: 10.1038/s41580-021-00407-0

[121] Su M, Feng G, Liu Z, Li Y, Wang R. Tapping on the black box: How is the scoring power of a machine-learning scoring function dependent on the training set? Journal of Chemical Information and Modeling. 2020;**60**(3):1122-1136. DOI: 10.1021/acs.jcim.9b00714

[122] Guedes IA, Barreto AM, Marinho D, Krempser E, Kuenemann MA, Sperandio O, et al. New machine learning and physics-based scoring functions for drug discovery. Scientific Reports. 2021;**11**(1):3198. DOI: 10.1038/s41598-021-82410-1

[123] Yang X, Wang Y, Byrne R, Schneider G, Yang S. Concepts of artificial intelligence for

computer-assisted drug discovery. Chemical Reviews. 2019;**119**(18):10520-10594. DOI: 10.1021/acs.chemrev.8b00728

[124] Chan HS, Shan H, Dahoun T, Vogel H, Yuan S. Advancing drug discovery via artificial intelligence. Trends in Pharmacological Sciences. 2019;**40**(8):592-604. DOI: 10.1016/j.tips.2019.06.004

[125] Bennett KP, Campbell C. Support vector machines: Hype or hallelujah? ACM SIGKDD Explorations Newsletter. 2000;**2**(2):1-3. DOI: 10.1145/380995.380999

[126] Vitali E, Gadioli D, Palermo G, Beccari A, Cavazzoni C, Silvano C. Exploiting OpenMP and OpenACC to accelerate a geometric approach to molecular docking in heterogeneous HPC nodes. The Journal of Supercomputing. 2019;**75**:3374-3396. DOI: 10.1007/s11227-019-02875-w

[127] Santos-Martins D, Solis-Vasquez L, Tillack AF, Sanner MF, Koch A, Forli S. Accelerating AutoDock4 with GPUs and gradient-based local search. Journal of Chemical Theory and Computation. 2021;**17**(2):1060-1073. DOI: 10.1021/acs.jctc.0c01006

[128] Wang J, Dokholyan NV. MedusaDock 2.0: Efficient and accurate protein–ligand docking with constraints. Journal of Chemical Information and Modeling. 2019;**59**(6):2509-2515. DOI: 10.1021/acs.jcim.8b00905

[129] Fan M, Wang J, Jiang H, Feng Y, Mahdavi M, Madduri K, et al. GPU-accelerated flexible molecular docking. The Journal of Physical Chemistry B. 2021;**125**(4):1049-1060. DOI: 10.1021/acs.jpcb.0c09051

[130] Bai Q, Tan S, Xu T, Liu H, Huang J, Yao X. MolAICal: A soft tool for 3D drug design of protein targets by artificial intelligence and classical algorithm. Briefings in Bioinformatics. 2021;**22**(3):bbaa161. DOI: 10.1093/bib/bbaa161

[131] Rifaioglu AS, Nalbat E, Atalay V, Martin MJ, Cetin-Atalay R, Doğan T. DEEPScreen: High performance drug–target interaction prediction with convolutional neural networks using 2-D structural compound representations. Chemical Science. 2020;**11**(9):2531-2557. DOI: 10.1039/C9SC03414E

[132] Yuan Y, Shi C, Zhao H. Machine learning-enabled genome mining and bioactivity prediction of natural products. ACS Synthetic Biology. 2023;**12**(9):2650-2662. DOI: 10.1021/acssynbio.3c00234

[133] Gaulton A, Hersey A, Nowotka M, Bento AP, Chambers J, Mendez D, et al. The ChEMBL database in 2017. Nucleic Acids Research. 2017;**45**(D1):D945-D954. DOI: 10.1093/nar/gkw1074

[134] Kwon Y, Park S, Lee J, Kang J, Lee HJ, Kim W. BEAR: A novel virtual screening method based on large-scale bioactivity data. Journal of Chemical Information and Modeling. 2023;**63**(5):1429-1437. DOI: 10.1021/acs.jcim.2c01300

[135] McGibbon M, Money-Kyrle S, Blay V, Houston DR. SCORCH: Improving structure-based virtual screening with machine learning classifiers, data augmentation, and uncertainty estimation. Journal of Advanced Research. 2023;**46**:135-147. DOI: 10.1016/j.jare.2022.07.001

[136] Zhang B, Li H, Yu K, Jin Z. Molecular docking-based computational platform for high-throughput virtual screening. CCF Transactions on High Performance Computing. 2022;**4**:63-74. DOI: 10.1007/s42514-021-00086-5

[137] Ohue M, Aoyama K, Akiyama Y. High-performance cloud computing for exhaustive protein–protein docking. In: Advances in Parallel & Distributed Processing, and Applications: Proceedings from PDPTA'20, CSC'20, MSV'20, and GCC'20. Springer International Publishing; 2021: 737-746 p. DOI: 10.1007/978-3-030-69984-0_53

[138] Lin Z, Zou J, Liu S, Peng C, Li Z, Wan X, et al. A cloud computing platform for scalable relative and absolute binding free energy predictions: New opportunities and challenges for drug discovery. Journal of Chemical Information and Modeling. 2021;**61**(6):2720-2732. DOI: 10.1021/acs.jcim.0c01329

Chapter 8

Applications of Molecular Docking Techniques in Repurposing of Drug

Ajay Kumar Shukla and Joohee Pradhan

Abstract

The applications of molecular docking techniques have played a key role in paradigm change in the field of drug development by providing a potent toolkit for the investigation. In addition, drug repurposing of already approved medications and for a novel treatment that was not previously recommended is known as drug repurposing. During the critical situation, it has attracted a great deal of attention. Molecular docking techniques have emerged as a necessity to expedite the drug development process and medication requirements. They promote a validated and cost- and time-effective method of creating novel pharmaceuticals. Molecular docking technologies facilitate the prediction of the binding interactions between small molecules and target proteins. Researchers can more thoroughly evaluate the potential efficacy of well-calibrated pharmaceuticals against novel disease targets. We will explain in this chapter how molecular docking was initially used to help with the drug discovery process. Next, we demonstrate the emerging and newer uses and applications of docking, such as target fishing and profiling, polypharmacology, drug repurposing, and adverse effect prediction. We also discussed about the potential of this technique, future applications, and its combination with other emerging techniques like artificial intelligence.

Keywords: molecular docking techniques, repurposing, paradigm, validation, prediction, polypharmacology

1. Introduction

Drug repurposing, also known as drug repositioning, drug reprofiling, indication expansion, or indication shift, is the process of finding new uses for pharmaceuticals that were previously approved but shelved, abandoned, or classified as experimental. While not a novel approach, drug repurposing has gained significant traction in the last 10 years: over one-third of recent approvals have been associated with this technique, and these drugs today account for approximately 25% of the pharmaceutical industry's yearly income [1].

1.1 Definition, extent, constraints, and scientific underpinnings of medication repurposing

Drug repurposing is the methodical process of discovering new uses for already approved medications, sometimes in tandem with their release into the general public and eventual generic availability. According to Langedijk et al. [2], this data was gathered by searching PubMed on November 21, 2019, for titles or abstracts that included the phrases "Drug repurposing" for the publication year 2018.

The original definition of drug repurposing has expanded to include not just medications that have been made generic but also active ingredients that were removed off the market because of safety concerns, or those that failed the clinical phase owing to toxicity or insufficient efficacy. It does not, however, include materials that have not yet undergone clinical testing, such as those kept in chemical libraries for screening by academic and commercial research teams. Hit-to-lead chemistry is usually required for compounds from such libraries in order to maximize their medicinal effects. Moreover, it is thought that the selective optimization of side activities (SOSA), as put out by Professor Wermuth, goes beyond repurposing. In SOSA, the biological characteristics that cause side effects are identified and made better through chemical alteration, making it possible to suggest the altered medication for a new use. This is not the same as repurposing, which spares the drug's structural changes [3].

Repurposing, on the other hand, makes use of two main scientific pillars: (1) the discovery that several diseases have similar biological targets, made possible by the decipherment of the human genome, and (2) the notion of pleiotropic medications. Drug repurposing requires an understanding of the complex interactions that exist between diseases, medicines, and targets. *In silico* techniques such as data mining, machine learning, ligand-based, and structure-based methodologies are essential for clarifying these interactions [4].

This talk highlights the current capacity to understand diseases from a molecular perspective, including a range of aspects such as genetic components, biomarkers, signaling pathways, and environmental factors that contribute to their pathogenicity. Determining the degree of similarity across diseases that share molecular characteristics has become crucial thanks to computational approaches, in particular data mining [5]. The discovery that 48 genes and four signaling pathways are shared by diseases like Parkinson's and Alzheimer's serves as an example of this paradigm shift Dovrolis et al. [6]. Finding protein targets that cross over into other diseases suggests that a single drug could be effective against a variety of pathological conditions.

The principal therapeutic actions and concurrent adverse effects of modern pharmacological substances are well-described. Drugs have a variety of effects due to their complex pleiotropic interactions with primary and secondary biological targets. This complexity raises the possibility of using medications created for one condition to treat another, provided that one of their secondary targets is present in the new illness [7]. Furthermore, the discovery of similar targets among diseases not only raises the prospect of a common pharmacological intervention but also opens the door to the creation of medications with a variety of purposefully designed side effects. Pan-kinase inhibitors are a good example of this idea in oncology, as their various yet coordinated actions work together to increase clinical efficacy.

The analogy between diseases and drugs extends to the realm of phenotypic analysis, disregarding therapeutic indications. By subjecting drugs with different therapeutic purposes to scrutiny based on phenotypic similarity, it becomes conceivable to discern potential efficacy across divergent indications. Thus, a high similarity score between two drugs prescribed for distinct diseases may signify their potential effectiveness in both scenarios [7]. This nuanced understanding of molecular and phenotypic intricacies not only broadens therapeutic possibilities but also underscores the potential for synergistic drug effects, elucidating a comprehensive approach to disease management and drug development.

1.2 Benefits of repurposing medications

Repurposing drugs has a number of related benefits. The main focus of these benefits is the simplification of the regulatory procedures involved in bringing a medicine that has already received approval onto the market, especially in countries like the United States. This strategy makes use of preexisting information, particularly on the drug's safety and toxicity, to speed up the first stages of repositioned drug development [8]. Consequently, this accelerated process not only reduces development timelines significantly but also incurs substantial cost savings, estimated to be over 80% according to Naylor and Schonfeld [9].

Moreover, the likelihood of successfully bringing a repurposed drug to market is notably enhanced, with a reported increase of 150% compared to the introduction of a novel drug, as highlighted by Thayer [10]. It is important to remember, though, that a drug's intended indication and the level of safety it requires are closely related. Tolerability of side effects is therefore correspondingly reduced when a medication is repositioned for a disease that is less severe or critical than its original prescription [11].

The alterations to the drug's composition, dosage, or mode of administration require a thorough reassessment of its safety profile in light of these changed circumstances. This effectively changes the repositioned medication into a new pharmaceutical product, highlighting the significance of careful inspection during this transitional stage.

2. Conventional drug discovery versus repurposing of drugs

The standard drug discovery process is characterized by five sequential stages, commencing with FDA review, followed by preclinical and discovery research, safety evaluation, and clinical investigation and concluding with post-market safety surveillance administered by the FDA. *De novo* identification of novel molecular entities (NME) is a crucial step in this approach. The protracted and resource-intensive nature of this approach heightens the likelihood of failure [12].

In contrast, drug repurposing involves a streamlined four-step progression: development, compound acquisition, compound identification, and FDA post-market safety monitoring, as illustrated in **Figure 1** [14]. The application of cheminformatics and bioinformatics technology, along with the readily available large biological and structural databases, has significantly reduced the time and cost commitment to drug development while also lowering the failure rate. Artificial intelligence (AI) technologies, structure-based drug design (SBDD), and *in silico* approaches are largely responsible for the modern acceleration of the drug repurposing process [14, 15].

Historically, the repurposing strategy using licensed drugs for new therapeutic indications has worked well, especially when it comes to coincidental discoveries. Compared to traditional drug discovery programs, which will be explained below, the

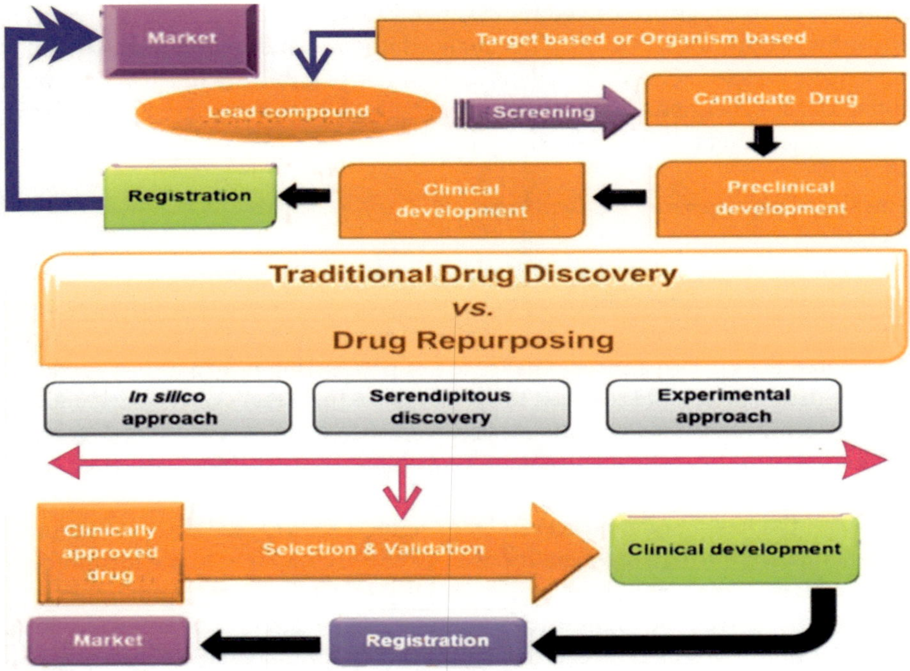

Figure 1.
Conventional drug discovery versus repurposing of drugs [13].

previously mentioned method seems to be rather helpful in the drug development domain. As an illustration, Pfizer repurposed sildenafil (Viagra), a phosphodiesterase-5 (PDE5) inhibitor, which was originally created to treat coronary artery disease (angina). This repurposing tactic may have helped save costs in addition to accelerating the development schedule. Phase II and III clinical trials are currently in progress, evaluating the feasibility of repurposing the oral antidiabetic medication metformin, commonly referred to as glucophage, as a treatment for cancer [16, 17].

Drug repurposing has various advantages over conventional drug development techniques. Interestingly, when compared to typical drug discovery programs, a significant decrease in the amount of time and money needed for research and development is seen. The conventional process for creating a new medication is estimated to span 10–16 years, whereas drug repurposing strategies may achieve this in a more time-efficient manner, ranging from 3 to 12 years. The financial implications are also noteworthy, with drug repurposing strategies costing as low as $1.6 billion, in stark contrast to the typical expenses associated with conventional drug development strategies, which can escalate to $12 billion. Furthermore, the identification of new pharmacological targets using drug repurposing strategies takes researchers only 1–2 years, with an average development period of 8 years for a repositioned medicine [18, 19].

The repurposing of pharmaceutical agents circumvents the customary 6–9 year timeline inherent in traditional drug development paradigms, facilitating a direct transition to preclinical evaluations and clinical trials. This departure mitigates cumulative risks, temporal expenditures, and financial outlays associated with drug development. Empirical evidence indicates that the approval timeline for repurposed drugs is notably shortened to approximately 3–12 years compared to conventional drug development, resulting in a concurrent reduction in development costs by 50–60%.

The initiation of a drug repurposing initiative is characterized by the preexistence of an array of preclinical data, including pharmacological and toxicological information, as well as clinical efficacy and safety data. This buildup is the result of the drug candidate's previous development through the early phases, which include structural optimization and early preclinical trialing. Additionally, it is possible that the proposed medication has received regulatory approval and, as a result, has a clearly established safety and clinical efficacy profile. Because of this, this method significantly lowers development costs while also reducing the dangers related to early-stage failures, which are common in traditional approaches. This dual benefit raises the probability of clinical safety and the overall success rate of medication development [20, 21].

In the inaugural phase of a repurposing development initiative, the utilization of preexisting data encompassing pharmacokinetics, toxicology, clinical trials, and safety profiles confers distinct advantages over the conventional drug discovery paradigm. These advantages materialize in the form of expedited development timelines, reduced developmental expenditures, and mitigated risks of clinical failure. The temporal requirements for the development of a repurposed pharmaceutical are estimated to span between 3 and 12 years, a noteworthy reduction compared to the 10–17 years typically associated with traditional drug discovery programs. This temporal efficiency is concomitant with a substantial decrease in associated costs. The average expenditure for bringing a novel drug to market through conventional means is approximately $1.24 billion, whereas drug repurposing incurs an outlay equivalent to or less than 60% of traditional drug discovery costs. Such economic efficiencies result in considerable savings of both time and capital for the repurposing entity.

Moreover, the strategic emphasis of conventional drug discovery predominantly revolves around identifying therapeutics for chronic and intricate diseases. On the other hand, the drug repurposing strategy focuses its efforts on creating medications that are specifically designed to treat infectious diseases that are quickly becoming more prevalent, as well as diseases that are resistant to treatment and neglected illnesses (NTDs). This change in emphasis is in line with how healthcare requirements are changing. In addition, the development of cheminformatics and bioinformatics techniques, along with the large databases and repositories of omics data (metabolomics, proteomics, transcriptomics, genomes, etc.), enhances the use of disease-targeted repurposing techniques. These methods make it easier to investigate hitherto unrecognized mechanisms of action, such as unknown pharmacological targets, latent drug-drug interactions, and new disease biomarkers. The effectiveness and efficiency of drug repurposing initiatives are increased by the incorporation of these cutting-edge strategies [20].

3. Techniques for repurposing drugs

As shown in **Figure 2**, there are two primary DR strategies: off-target and on-target. When a medication molecule's established pharmacological mechanism is applied to a novel therapeutic indication, on-target drug delivery occurs. This approach targets a different disease with the therapeutic chemical using the same biological target [16, 17].

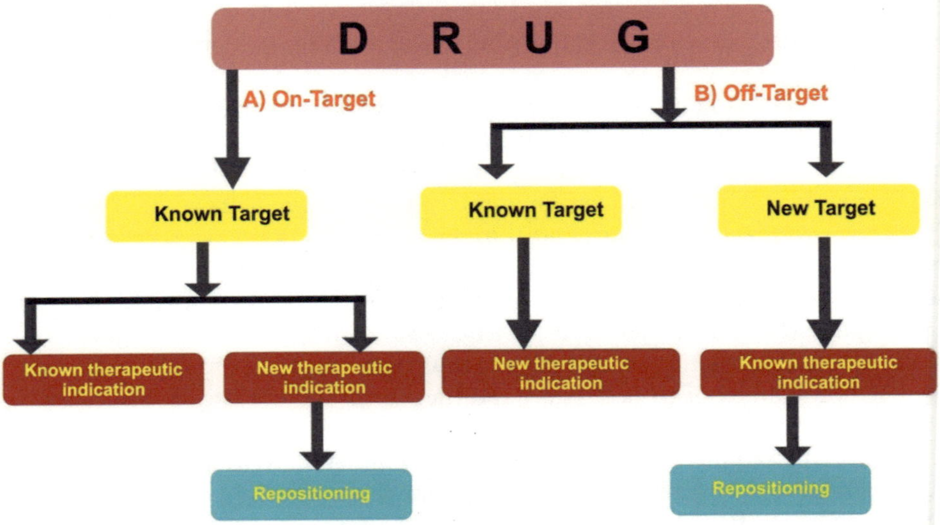

Figure 2.
Drug repurposing tactics: on- and off-target [13].

During the repurposing process, minoxidil (Rogaine) exhibits a pronounced on-target profile in which the medication acts on a same molecular target to produce two different therapeutic effects. Minoxidil was developed as an antihypertensive vasodilator before evolving into a medication for hair loss. Minoxidil acts as an antihypertensive vasodilator by modulating potassium channels and exhibiting vasodilatory qualities. This allows for increased blood, oxygen, and nutrition supply to hair follicles. The effectiveness of this medication in treating male pattern baldness (androgenic alopecia) is supported by its pharmacological activities. On the other hand, the off-target profile's pharmacological mechanism is still unknown. New therapeutic indications are being explored as a result of interactions between pharmaceuticals and drug candidates that are repurposing and novel targets outside of their initial scope. Thus, in the context of relocated minoxidil, both the molecular targets and therapeutic indications reflect new aspects [22].

The drug aspirin (Colsprin) is a great example of an off-target profile. Aspirin, which functions as a nonsteroidal anti-inflammatory medication (NSAID), has long been used to treat a variety of inflammatory and pain-related ailments. Additionally, it suppresses blood coagulation and acts as an antiplatelet drug by inhibiting the normal function of platelets. As a result, aspirin is useful in the treatment of heart attacks and strokes. Additionally, recent research has hinted at a potential treatment for prostate cancer based on aspirin.

4. Approaches of drug repurposing

Aspirin is a notable example of an off-target profile design (Colsprin). As a nonsteroidal anti-inflammatory medication (NSAID), aspirin has been used traditionally to treat a range of inflammatory and nociceptive disorders. It also inhibits the intrinsic properties of platelets, which prevents blood coagulation (clot formation), which is why it is categorized as an antiplatelet drug. As a result, aspirin is utilized in the treatment of myocardial infarctions and cerebrovascular accidents. Additionally, recent publications have demonstrated an innovative use of aspirin in the treatment of prostate cancer [23, 24].

On the other hand, the *in silico* repurposing methodology makes use of cheminformatics, bioinformatics, and computational biology approaches to perform virtual screening of large drug/chemical libraries that are available in public databases. This method depends on identifying putative bioactive chemicals based on how the drug molecule and the protein target interact chemically (**Table 1**) [25].

Since it has shown such great promise in drug development efforts, the *in silico* methodology has become more and more popular in recent decades. With regard to the chemical structures of proteins, pharmacophore models, and bioactive chemicals, this approach makes use of the abundance of publically available data. Numerous pharmaceutical enterprises and research laboratories dedicated to drug discovery have successfully incorporated *in silico* tools and techniques, particularly for exploring diverse chemical spaces in the quest for novel therapeutic agents.

In comparison with experimental-based approaches, *in silico* repurposing offers several advantages, including a diminished failure rate, accelerated development timelines, and cost-effectiveness. However, a notable limitation of this approach is its dependency on precise structural information pertaining to drug targets. In instances where such information is unavailable, reliance on medication phenotypic or genotypic profiles relevant to the associated disease becomes imperative, as highlighted by Rosa and Santos [27]. The presented figure serves to symbolize various strategies employed in medication repurposing endeavors (**Figure 3**).

Recently, researchers and scientists have adopted an integrated approach that combines *in silico* and empirical methodologies in the quest to discover novel therapeutic indications for medications that have previously been approved for use. In this integrated model, clinical studies and preclinical biology experiments including both *in vitro* and *in vivo* evaluations combine to evaluate the results of computational approaches. Demonstrating superior efficacy compared to fortuitous discoveries, the concurrent and systematic application of computational and experimental modalities imparts a dependable and rational foundation for the exploration of new therapeutic applications. This hybridized strategy further affords opportunities for expeditious and streamlined repurposing of medications, establishing itself as a dependable and credible approach in pharmaceutical development [28].

Activity-based approach	***In silico*-based approach**
In vitro and *in vivo* experimental screening	Virtual or computational screening
Assays for screening cells/organisms and targets	Protein target-oriented screening
Requires neither drug-induced cell/disease phenotypic data nor target protein structural knowledge	Needs knowledge of the target proteins' structures as well as drug-induced cell or disease phenotypic data
Labor-intensive and time-consuming	Labor and time efficiency
Reduced false-positive hit rate during the screening process	Greater percentage of false-positive results while screening

Table 1.
Disparities between in silico- and activity-based methods for medication repurposing [13, 25, 26].

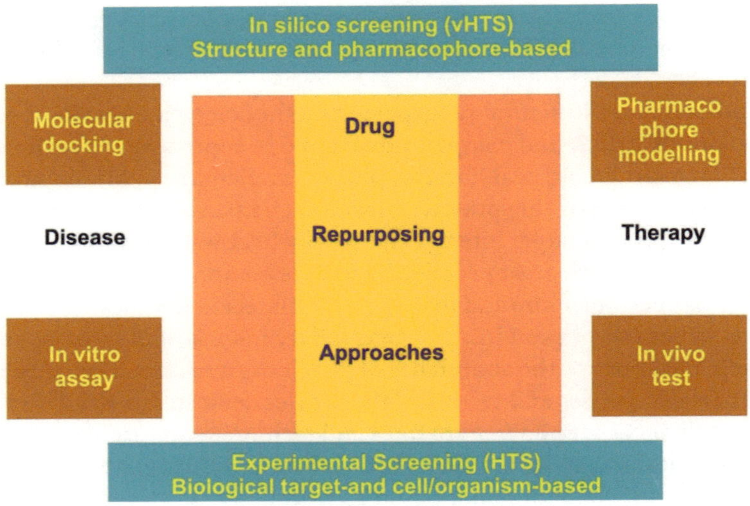

Figure 3.
Approaches of drug repurposing [13].

5. Methodologies of drug repurposing

Drug repurposing techniques can be roughly classified into three classes based on the quality and depth of available data on biological activity, pharmacology, and toxicology. The three main categories in which these approaches fall are (i) drug-centric, (ii) target-centric, and (iii) disease/therapy-centric.

The structural characteristics of therapeutic molecules, biological activity, side effects, and toxicities are all methodically evaluated in the drug-centric approach. This methodology seeks to discover substances with unique biological effects by means of experiments carried out on animals or cells. This repurposing paradigm emerged from studies that were largely focused on evaluating the biological efficacy of therapeutic compounds, frequently without a complete understanding of the underlying biological targets. Its roots are in conventional pharmacology and drug development concepts. The discovery of sildenafil through chance or clinical observation is one of the many noteworthy achievements in drug repurposing that the drug-centric strategy has historically produced [29]. This approach demonstrates how well it works to find new therapeutic uses for medications that already exist by using thorough evaluations of their pharmacological and biological characteristics.

Target-based methodology comprises high-throughput and/or high-content screening (HTS/HCS) of drugs against a target protein molecule or biomarker *in vitro* and *in vivo*, followed by virtual high-throughput screening (vHTS) or *in silico* screening of drugs or compounds from drug libraries/compound databases, such as molecular docking or ligand-based screening. This strategy has a substantially greater success rate in drug development than drug-oriented methods since most biological targets reflect the pathways or mechanisms causing disease [30].

In cases where the disease model has access to more data, the illness/therapy-oriented paradigm is appropriate in disease-research. In this case, decisions about disease-specific treatments can be informed by proteomics (disease-specific target proteins), genomics (disease-specific genetic information), metabolomics

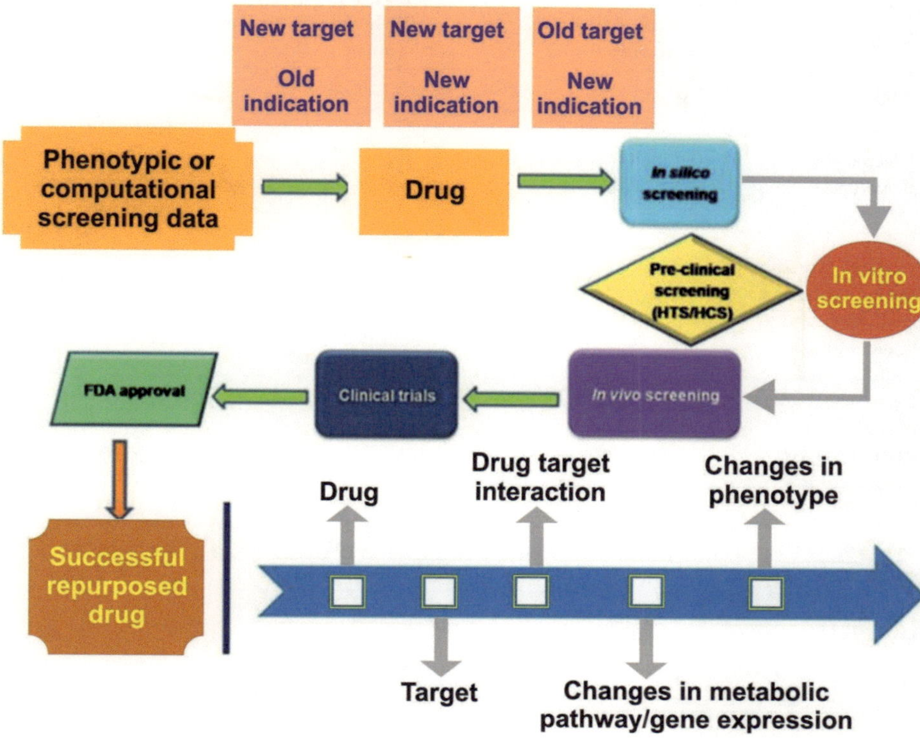

Figure 4.
The procedures and techniques used in medication repurposing [13].

(disease-specific metabolic pathways/profile), and phenotypic data (disease-specific disease processes, pharmacological targets, off-target mechanisms, pathological conditions, adverse and side effects, etc.). Consequently, it requires the construction of discrete disease networks, the discovery of genetic expression, the evaluation of significant targets, the identification of disease-causing protein molecules linked to cell and metabolic processes of interest in the disease model, and so on. **Figure 4** illustrates the steps and procedures required in repurposing medications.

The production of tiny medicinal compounds and biologics approved by the FDA is mostly accomplished through the application of target-based and drug-based phenotypic screening methods. Phenotypic drug screening methods identify therapeutic possibilities in small molecule libraries by means of fortuitous findings. Target-based strategies look for novel drugs by utilizing known target molecules. The treatment/therapy-based repurposing methodology and the disease-based strategies are comparable and presented in **Table 2** [31].

The following is a brief description of several of the available repurposing techniques shown in the **Table 2**.

5.1 Biological tests and experimental screens

Directed toward certain illness models and medications are used in blinded search or screening methods, whereby results are discovered by chance. These approaches' greater flexibility in screening a wide range of medications or illnesses is one of its advantages.

Techniques	Method type and category	Technique/particular strategy	Examples
Drugs			
Screening by phenotype	Target-based and blinded screening	HTS/HCS screening both *in vivo* and *in vitro*	For erectile dysfunction, sildenafil; for breast cancer, rituximab
Target 3D structure, chemical structure, medication, and ligand information	Using targets, cheminformatics	Ligand-based screening, fragment-based screening, molecular docking, and *in silico* screening	Etoposide (bladder cancer), fluorouracil (lung cancer)
Details about drug targets, chemical structure, target, and medication information	Based on knowledge, cheminformatics and Bioinformatics	Prediction of drug targets	Ketoconazole and simvastatin for breast cancer
Details of clinical trials and side effects	Based on knowledge, bioinformatics	Studies of drug similarity	—
FDA-approved labeling	Based on knowledge, bioinformatics	Studies of drug similarity	—
Disease-oriented			
Information on the available pathway	Based on knowledge, bioinformatics	Finding the cause of the sickness and addressing the main objectives	Vismodegib (cancerous skin)
Data on disease omics and genetics	Bioinformatics based on signatures	Using genomics and gene signatures to find important targets	—
Protein interaction network, accessible pathway information, and disease omics data	Network biology is either pathway- or network-based.	Examining disease-specific networks and pathways to find important targets	Dasatinib with sunitinib (brain tumor, breast cancer)
Therapeutic in nature			
Data on drug omics	Bioinformatics and/ or network biology, signature-based or signature-and-network based	Examining gene signatures	Acute lymphoblastic leukemia (Sirolimus); neurological diseases (Fasudil)
Data on medication and disease omics	Bioinformatics based on signatures	Drugs and illnesses share similarities	Lung cancer: cetirizine; inflammatory bowel disease: topiramate
Protein interaction network, disease pathway, and drug omics data	Systems biology, network biology, and targeted-mechanism biology	Clarifying specific routes	Clomifene and daunorubicin (breast cancer)

Table 2.
A thorough summary of the various methods for repurposing medications along with relevant examples. A few currently in use medication repurposing methods [13, 31, 32].

5.2 Target-based methods include

Ligand-based screening, molecular docking, and high-throughput and/or high-content screening (HTS/HCS) of therapeutic molecules for a target protein or biomarker of interest. They also do *in silico* screening of compounds or pharmaceuticals from huge chemical libraries. These strategies have a higher likelihood of finding drugs or drug leads than blinded search techniques. Furthermore, the entire screening process takes less time to complete.

5.3 Knowledge-based approaches

Use cheminformatics or bioinformatics procedures to gather data from clinical trials on drug profiles, chemical structures of targets and pharmaceuticals, drug-target networks, side effects, signaling, and metabolic pathways. The information level of knowledge-based approaches is sufficiently rich in comparison with target-based or blindfolded methods. Since the existing data may be predicted, it can also be used to anticipate novel and unknown mechanisms, such as novel disease biomarkers; medication targets that are not yet identified; or similarities between drugs.

5.4 Gene signatures created

From disease omics data (genomics data), with or without medicines, are used in signature-based techniques to identify unknown off-targets or unknown sickness processes. Genomic data can be accessed through public databases. One advantage of these methods is that they can be used to look into unknown drugs' mechanisms of action. Unlike knowledge-based approaches, signature-based approaches employ computational techniques to investigate pharmacological processes at the molecular level, such as variations in gene expression.

5.5 Pathway or network-based methods

Is made possible by route- or network-based approaches, which make use of strategies based on network analysis or pathway informatics. These methods make use of accessible signaling or metabolic pathways, protein interaction networks, and disease-omics data. One intrinsic benefit of these approaches is their ability to concentrate on a limited subset of proteins inside a particular network, as opposed to addressing a significant number of proteins distributed across several target molecules or signaling networks.

5.6 Targeted mechanism-based approaches

Integrate treatment omics data, existing knowledge of signaling circuits, and protein interaction networks to define the undiscovered mechanisms of action of drugs. In addition to being used to identify the processes connected to medications or illnesses, these techniques also have the advantage of being able to identify the mechanisms directly linked to the drug therapies for certain diseases [31–33].

6. Polypharmacology

Polypharmacology denotes the capacity of a pharmaceutical agent or compound to engage with multiple molecular targets within the physiological milieu. Conventional paradigms of drug discovery have historically emphasized the development of compounds that exhibit specific interactions with individual targets to elicit therapeutic effects. However, numerous pathological conditions manifest intricate biological networks and pathways, necessitating a shift toward interventions that concurrently target multiple components, thereby potentially augmenting therapeutic efficacy.

Molecular docking stands as a computational methodology integral to the realm of drug discovery and design. This technique entails the computational prediction of the optimal orientation and conformation of a diminutive molecule (referred to as a ligand) upon binding to a target macromolecule, typically a protein, resulting in the formation of a stable complex. The elucidation of these molecular interactions is pivotal for comprehending the binding dynamics between a pharmaceutical agent and its designated target, thereby facilitating the refinement of drug candidates to optimize efficacy and selectivity. The interconnection between polypharmacology and molecular docking within the drug discovery process is manifested in the following symbiotic relationship:

6.1 Identification of multiple targets

Within the realm of polypharmacology, investigators endeavor to discern and comprehend the interactions exhibited by a pharmaceutical agent with numerous targets implicated in a disease pathway. Molecular docking serves as a predictive tool for determining the binding affinity and mode of interaction between a ligand and various target proteins. The resultant information holds significance in the evaluation of a drug's potential to engage with multiple targets.

6.2 Optimizing polypharmacological agents

Computational methodologies, inclusive of molecular docking, play a pivotal role in formulating and refining polypharmacological agents. Through an understanding of molecular-level binding interactions, researchers can modify drug candidates to enhance their affinity for multiple targets.

6.3 Virtual screening for polypharmacological agents

Molecular docking frequently features in virtual screening processes, wherein extensive databases of chemical compounds are computationally sifted to discern potential drug candidates. Virtual screening can be customized to pinpoint compounds capable of interacting with multiple targets, thereby fostering the discovery of polypharmacological agents.

6.4 Understanding structure-activity relationships (SAR)

Molecular docking significantly contributes to Structure-Activity Relationship (SAR) studies by furnishing insights into how alterations in the chemical structure of a compound may impact its binding to diverse targets. SAR information assumes a pivotal role in optimizing drug candidates with polypharmacological potential.

6.5 Network pharmacology

Polypharmacology aligns closely with the concept of network pharmacology, which scrutinizes interactions among multiple targets and pathways within a biological network. Molecular docking, coupled with other computational methodologies, aids in the analysis of network pharmacology by forecasting and validating interactions within a complex biological system.

Molecular docking emerges as an invaluable instrument in the investigation and refinement of polypharmacological agents, affording insights into the molecular intricacies of interactions between drugs and their targets. This approach holds particular relevance in the context of intricate diseases where modulation of multiple targets is imperative for achieving efficacious therapeutic outcomes [32, 33].

7. The procedure of repurposing drugs

Medication repurposing is the methodical process of identifying and exploring new therapeutic uses for pharmaceuticals that already have a market. By exploring the possibility of using previously approved or experimental medications to treat conditions other than the ones for which they were initially approved, this novel strategy avoids the traditional method of *de novo* drug development. Adopting drug repurposing techniques could provide advantages in terms of timing and resources when compared to the traditional drug discovery process.

Molecular docking, which forecasts the binding affinities and spatial orientation of a small molecule, such as a possible medication candidate, with a target macromolecule, such as a protein, is one of the most significant computational techniques in drug discovery. In the subject of drug repurposing, molecular docking is essential because it might reveal possible interactions between current drugs and new therapeutic targets. The processes that are frequently used in medication repurposing are outlined here in an organized procedural overview that includes the function of molecular docking.

7.1 Target identification

Target identification aims to ascertain a fresh therapeutic target affiliated with a designated disease. This target may manifest as a protein, enzyme, or other molecular constituent participating in the pathogenesis of the disease.

7.2 Drug selection

This involves selection of established drugs anticipated to interact with the newly discerned target. These drugs may originate from diverse reservoirs, comprising approved drugs for disparate indications, investigational drugs, or previously unsuccessful drugs exhibiting promise against an alternative target.

7.3 Molecular docking

To simulate how the selected medications would interact with the target protein, molecular docking tools are used. To do this, the medications' affinities and binding mechanisms with the target must be predicted.

7.4 Scoring and analysis

The outcomes of the docking simulations are assessed using scoring functions to prioritize potential interactions. Elevated scores signify heightened binding affinities. Scrutinize the anticipated binding modes to elucidate the nature of drug-target interactions.

7.5 Validation

Validation through experimental means is imperative to affirm the forecasted interactions. This may involve *in vitro* investigations, such as biochemical assays, alongside *in vivo* studies gauging therapeutic efficacy in animal models.

7.6 Clinical testing

Clinical trials are conducted to evaluate the safety and efficacy of the repurposed medicine for its novel indication when preclinical findings show promise.

The advantages intrinsic to drug repurposing encompass the prospect of an expedited development timeline, diminished developmental expenditures, and the exploitation of preexisting safety and pharmacokinetic data. Nonetheless, challenges persist, including the procurement of intellectual property rights for novel applications and the potential manifestation of off-target effects.

Molecular docking, as a computational methodology, contributes to the preliminary screening and identification of conceivable drug-target interactions, facilitating the prioritization of compounds for subsequent experimental validation. It is imperative to underscore that while molecular docking affords invaluable insights, experimental validation remains paramount to substantiate the predicted interactions, ensuring both the efficacy and safety of repurposed drugs [25].

7.7 Drug repurposing opportunities

7.7.1 Uncommon and untreated illnesses

Repurposing pharmaceuticals is a compelling strategy, particularly when addressing rare and underdiagnosed illnesses that have unfavorable economic dynamics for the development of new medications. Under these conditions, the role of academic and nonprofit organizations becomes more significant in guiding the drug discovery process, supported by public policies and some regulatory actions intended to encourage research in these disease areas. These policies include financial awards, accelerated approval procedures, tax exemptions, and waivers of regulatory fees. Repurposed medications make up a significant percentage of the Drug for Neglected Diseases initiative (DNDi) clinical trial portfolio. Examples of these include miltefosine, fexinidazole, fosravuconazole, and AmbisomeTM. Fexinidazole is noteworthy for being the first medication developed for advanced-stage sleeping sickness to be taken orally in the previous three decades. DNDi was able to develop it for just USD 62.5 million, which is a small amount compared to the estimated $1 to $3 billion that is usually involved in *de novo* drug development [34].

Crucially, some of the obstacles that arise in commercial settings like worries about off-patent medications, for example, become less formidable when it comes to finding treatments for undertreated ailments. The fundamental motivation for research into treatment solutions for these kinds of illnesses is not profit. As a result, efforts may be made to repurpose inexpensive, off-patent medications in order to guarantee widespread accessibility. Computational methods for predictive repurposing provide quick routes to testable ideas that can be translated into clinical applications in the field of uncommon diseases, which are frequently characterized by poorly known pathophysiology. Determining the genetic variants causing these ailments is made possible by large-scale genome sequencing projects, which speed up the development of medications that target the relevant proteins [35].

7.7.2 Precision medicine

Precision medicine represents an evolving paradigm that takes into account the individualized variations in genetic makeup, environmental exposures, and lifestyle factors to inform the selection or pursuit of tailored therapeutic interventions [36]. It is becoming clearer that several illnesses that were once thought to be solitary conditions really present as a spectrum of illnesses. Finding safer and/or more effective drugs is possible if they are designed to target particular subtypes within a larger pathological disease, or if they are matched to an individual's unique genomic, transcriptomic, proteomic, and metabolomic differences. The field of cancer is seeing a growing number of practical benefits from precision medicine. Exemplified by a recent case study of a patient with advanced colorectal cancer who was resistant to traditional chemotherapy and radiation and was exhibiting treatment-related damage [37], it emphasizes how important it is to repurpose drugs. To determine the expression of the V600E mutant BRAF protein and the MMR proteins (MSH2, MSH6, PMS2, and MLH1), the patient underwent immunohistochemistry investigation. Comprehensive genomic analysis revealed over 2000 genomic alterations, including point mutations, insertions, deletions, and copy-number variations. This included whole-genome sequencing of the pretreatment tumor and blood as well as whole transcriptome and whole genome sequencing of the metastatic tumor.

Genes that were differently expressed were found using transcriptome analysis; they included members of the proto-oncogene families FOS and JUN. These results validated the theory that renin-angiotensin system blockade can have therapeutic advantages. As a result, the patient experienced a dramatic and long-lasting response when the hypertension angiotensin II receptor antagonist irbesartan was repurposed as an anticancer medication. This is an example of how precision medicine can find new treatment approaches that are specific to the molecular details of each case.

7.7.3 Systems biology

An integrative viewpoint offered by systems medicine and network pharmacology challenges earlier drug discovery paradigms, particularly the target- and phenotypic-oriented methods connected to "rational" drug discovery. By employing network and metabolic control analysis as useful instruments, these approaches aid in the development of multi-target treatments or the choice of complementary drug combinations. The nifurtimox eflornithine combination therapy, which has been approved for inclusion in the World Health Organization's Model List of Essential Medicines for the treatment of advanced stages of the Gambiense form of sleeping sickness, provides a clear illustration of how this strategy can be used in practice.

Comparing this specific combination therapy to eflornithine monotherapy, there are benefits such as ease of administration, a shorter course of treatment, and possible defense against the development of resistant parasites. Interestingly, eflornithine, which was first developed for the treatment of cancer in the late 1970s, and nifurtimox, which was first approved for the treatment of American trypanosomiasis, have both undergone repurposing. Repurposing well-known medications as combination therapy is a well-trod path within the profession; the concept of combination drug repurposing adds another layer. Compared to single-drug therapy, the field of possible combination drug repurposing is far more expansive and difficult to exhaust due to the naturally combinatory nature of polypharmacy. Furthermore, when novel active compounds are identified, their potency is frequently found to be limited, which makes it difficult to use them in clinical settings right away since tolerated plasma drug concentrations are below limits that are effective. As an alternative to monotherapies, synergistic medication combinations have the potential to reduce the necessary therapeutic doses, which could increase the success rate of drug repurposing.

7.7.4 Cooperative models

Pharmaceutical corporations and academic institutions are beginning to realize how well they may work together to find new opportunities for repurposing existing products. Pharmaceutical companies have vast and significant chemical library warehouses that are frequently neglected and contain chemicals from failed or terminated medication prospects. These businesses also contribute to a great deal of experience in clinical development and translational research. Additionally, they provide access to screening technologies, which are generally difficult for most academic institutions to acquire and maintain. On the other hand, biotechnology companies and academic institutions make substantial contributions to the understanding of new fields in disease biology, which prepares the way for the creation of very creative drugs. Non-negligible benefits result from collaboration between these sectors, such as the development of human capacity through exchange of knowledge and training [20].

Potential approaches from the standpoint of intellectual property include investigating patent pools, implementing open licensing for the development of drugs targeted at underserved or uncommon diseases, and assisting academic institutions and staff in becoming patent owners of novel medical applications. To close the gap between stakeholders, innovative business models and collaboration are emerging. These models include financial sources like public money, venture capital, and nonprofit organizations. Certain medical fields, especially those related to uncommon illnesses, where medication repurposing plays a significant role, could benefit greatly from the application of these models.

8. Drug repurposing opportunities

After being approved by the United States Food and Drug Administration (US FDA) in 1998, sildenafil was brought to the pharmaceutical market and quickly became popular, according to Boolell et al. [38]. In a parallel setting, the well-known medication thalidomide experienced a reorientation. Thalidomide was first developed as a sedative by the German pharmaceutical company Grünenthal in 1957. However when it was given to expectant mothers to treat morning sickness, it was shown to have unanticipated teratogenic consequences. The ensuing discovery of severe birth abnormalities led to its discontinuation from distribution after being utilized in 46 nations, resulting in over 10,000 cases of limb and extremity deformities in newborns, with more than half succumbing within months of birth [39].

Over subsequent decades, research endeavors elucidated thalidomide's unanticipated anticancer properties. D'Amato and Folkman [40] observed its angiogenesis-inhibiting capabilities in animal experiments, opening avenues for further exploration. Further research, as demonstrated in studies [41, 42], revealed possible treatment advantages for multiple myeloma that was refractory and metastatic prostate cancer. These results culminated in the US FDA approving thalidomide and dexamethasone together for the treatment of multiple myeloma in 2006. Examples of many repositioned drugs that have been produced or are being developed from various FDA-approved or marketed drugs as well as investigational new drugs (IND) are displayed in **Table 3**. A few repositioned drugs that are now participating in COVID-19 clinical trials are also included.

Many datasets and software are publicly available for use in pathway analysis, proteomics, metabolomics, and genomics research. Numerous computational algorithms have previously been developed to expedite and streamline the repurposing process. A few notable databases utilized in pharmaceutical repurposing research are shown in **Table 4**.

9. Challenges associated with drug repurposing

Repositioners face significant challenges primarily stemming from the comparatively inadequate intellectual property safeguards applied to pharmaceuticals within this category. Such limitations can adversely impact the return on investment and act as deterrents for businesses engaged in their development [43]. Interestingly, repurposing drugs with the same active ingredient provides protection only in the form of a fresh application patent, which may be enhanced by a different formulation technique. This need results from the medication having previously been patented as a unique chemical entity. It is essential to recognize that the range of medicinal applications covered by application patents is systematically smaller than that of new chemical entities. This means that repositioners might have trouble stopping the off-label prescription of generic versions for applications that are patented. Additionally, the patents' legal standing may be weakened, especially in the event of any legal challenges claiming that the new indication was predicted based on scientific literature [11].

Notwithstanding these difficulties, companies that offer medications for orphan diseases that are defined in Europe as having a prevalence of little more than 5 in 10,000 may benefit more from them than from the relative drawbacks of patent protection. These benefits include a guaranteed duration of market exclusivity and fee reductions [44]. Notably, some companies, exemplified by Apteeus in France, specialize in "personalized" medication repurposing. In this approach, potential treatments are screened using patient cells originating from orphan diseases. This tailored strategy represents an innovative avenue within the repurposing landscape.

9.1 Difficulties in repurposing drugs

Repurposing drugs is promoted as a resource-efficient way to quickly produce new drugs. But oftentimes, academic researchers fail to fully address the necessary

Pharmacological category of drug	Uses	Novel uses	Developmental state
Ampicillin B (AMB), an antifungal medication	Fungus-related illnesses	The leishmaniasis	Developed already
NSAIDs and aspirin	Inflammation and pain	Antiplatelet CVDs	Developed already
		Prostate cancer	In progress
Antihistaminic astemizole	An allergic reaction, like urticaria	Malaria	In progress
Antidepressant atomoxetine	Depression	Hyperactivity and attention deficit	Developed already
Avermectin, a chelating agent	River-blindness and elephantiasis	Tuberculosis	In progress
Antibacterial antibiotic azithromycin	Infections with bacteria	COVID-19	In progress
Antidepressants, SSRIs, and bupropion	Depression	Giving up smoking	Developed already
NSAIDs, COX-2 inhibitors, and celecoxib	Inflammatory response	Colon and breast cancer	In progress
H2 receptor antagonist cimetidine	Stomach sore	Prostate, lung, and breast cancer	In progress
As an ALK inhibitor, crizotinib	Cancer (lymphoma; ALCL clinical study ongoing)	NSCLC	Developed already
Anti-inflammatory drug colchicine	Gout (arthritis gouty)	The pericarditis	Developed already
		COVID-19	In progress
Cardiotonic and Digoxin	CVDs, including heart attacks	Cancer of the Prostate	In progress
Acetaldehyde dehydrogenase inhibitor, disulfiram	Persistent alcoholism	Cancer	In progress
The immunosuppressant everolimus	suppressor of the immune system	Neuroendocrine tumors of the pancreas	Developed already
Antimetabolite, anticancer, and fluorouracil	Cancer	Breast cancer	Developed already
Antidepressant Fluoxetine	Depression	Dysphoria before menstruation	Developed already

Table 3.
Examples of pharmacological category of drug, uses, novel uses, and stages [13].

factors to guarantee that a medicine that has been repurposed is appropriate for a new application. The following discussion explores a number of issues related to drug repurposing.

9.2 Intellectual property and economic considerations

Legal obstacles impacting the patentability of new medical uses and/or the enforcement of patent rights may reduce the incentives for drug repurposing. While

Information Type	Database	Website
Chemical Structure	PubChem	http://pubchem.ncbi.nlm.nih.gov
	DrugBank	http://www.drugbank.ca/
	ChemSpider	http://www.chemspider.com
	ChemDB	http://www.chemdb.com
	Therapeutic Target Database(TTD)	http://bidd.nus.edu.sg/group/cittd/
Target 3D Structure	RCSB Protein Data Bank (PDB)	http://www.rcsb.org
	OCA	http://oca.weizmann.ac.il/oca-bin/ocamain
	Proteopedia	http://proteopedia.org
Drug-Target Information	Drugbank	http://www.drugbank.ca/
	Therapeutic Target Database (TTD)	http://bidd.nus.edu.sg/group/cittd/
	Pharmacogenetics Knowledge Base (PharmGKB)	http://www.pharmgkb.org/
	DrugMap Central (DMC)	http://r2g2drug.org/index.html
Protein Interaction Information	Human Protein Reference Database (HPRD)	http://www.hprd.org/
	Biological General Repository for Interaction	http://thebiogrid.org/
	Database of Interacting Proteins (DIP)	http://dip.doe-mbi.ucla.edu/dip/Main.cgi
	STRING	http://string-db.org/
Pathway Information	NCI Pathway Interaction Database (NCI-PID)	http://pid.nci.nih.gov/
	Kyoto Encyclopedia of Genes and Genomes (KEGG)	http://www.genome.jp/kegg/
	PathwayCommons	http://www.pathwaycommons.org/about/
Clinical Trial Information and Adverse Effects	Clinicaltrial.gov	http://clinicaltrials.gov
	AdverseReaction Database (Canada)	http://www.fda.gov/Drugs/
	SIDER	http://sideeffects.embl.de/
FDA Label Information	FDALABEL (US FDA)	http://www.fda.gov/ScienceResearch/
	DailyMed (US FDA)	http://dailymed.nlm.nih.gov/dailymed/about.cfm
	Structured Product Labeling (SPL)	http://www.dailymed.nlm.nih.gov/dilymed/about.cfm
Omics Data (Target/Drug)	NCBI-GEO	http://www.fda.gov/ForIndustry/DataStandards/
	Sequence Read Archive (SRA)	http://www.ncbi.nlm.nih.gov/geo/
	ArrayExpress	http://www.ncbi.nlm.nih.gov/Traces/sra/
	Cancer Cell Line Encyclopedia (CCLE)	http://www.ebi.ac.uk/arrayexpress/
	Sequence Read Archive (SRA)	http://www.broadinstitute.org/ccle/home

Table 4.
Information's, database, novel websites [13].

major pharmaceutical markets often allow patents for repurposed medicinal uses, several country laws initially present barriers to getting a patent for additional or secondary medical use. Furthermore, a plethora of potential off-label, unregistered uses for repurposing have been documented or are in clinical practice, despite the absence of controlled clinical trials endorsing such applications [35]. This existing knowledge, available to the public, may influence patentability and impede innovation. While it is feasible to secure a patent for a new indication for an off-patent medicine, challenges may arise if the new indication employs strengths and dosage forms already on the market. It may be necessary to use non-marketed strengths (ideally lower than those that were previously available) or a completely new formulation in order to fully maximize the benefits of drug repurposing [35, 45]. It is inappropriate to use more modern derivatives since they depart from the repurposing paradigm. In general, the European Union grants 2 years of market exclusivity and 8 years of data protection. If a second indication is created during the eight-year data exclusivity term, an extra year may be granted. On the other hand, the United States first provides 5 years, which can be renewed for an additional 3 years [46]. But these long times might not be favorable for achieving a sufficient return on investment; hence, different kinds of financial incentives are necessary to make drug repurposing profitable.

9.3 Availability of data and compounds

In the realm of drug development, the open-source paradigm is gradually gaining prominence, as noted in Ref. [20]. However, certain impediments persist, hindering public access to crucial data types, notably clinical trials. Accessibility concerns notwithstanding, challenges arise with data types such as image data, due to either their incompatibility with data mining, integration, or manipulation, or irregular provision [35]. The integration of disparate data types to enhance analytical capabilities proves computationally intensive [47].

The reluctance of certain pharmaceutical companies to disclose their chemical libraries, particularly failed drugs, poses challenges in diversifying potential applications and hinders drug repurposing efforts. Selectivity in partnership choices may create a fundamental obstacle, especially when a repurposed indication lies beyond the company's core disease area. Despite the potential value of shelved drugs as idle capital, missed opportunities, or postponed endeavors, the release of chemical libraries remains constrained. Effective facilitation and flexibility in administrative procedures are pivotal for compound distribution and material transfer agreement signatories. This holds true even when larger pharmaceutical companies express openness to crowdsourcing or collaboration with smaller entities such as boutique firms or academic groups. Complications may arise in cases involving compounds with generic active pharmaceutical components, particularly if the chemical is no longer available in the global market. The identification of a reliable vendor under such circumstances may prove challenging [47].

9.4 The area under consideration for drug repurposing

It is possible to argue that systematic efforts in medication repurposing could quickly exhaust the opportunities for repurposing pharmaceuticals for a particular ailment, despite the wide range of diseases requiring improved treatment approaches. The pool of candidates for repurposing is naturally small and grows yearly at a slow pace. For instance, a number of high-throughput screenings aimed at finding repurposed drugs with trypanocidal qualities for possible use in the treatment of Chagas disease have been observed over time. But these screenings frequently ignore the results of earlier low-throughput screenings [48, 49] and are directed toward specific pharmacological targets, requiring subsequent experimental validation through *in silico* or wet screens (see, for instance, [50, 51]). An essential concern arises regarding the appropriateness of further repurposed-oriented phenotypic screenings focused on Trypanosoma cruzi. Similarly, one may question whether interest in drug repurposing would wane over time with the execution of a series of systematic screenings on established drug collections.

A quick solution to this conundrum could entail changing strategy to focus on unnamed target ailments. Examining medication combinations instead of possibly repurposed monotherapies may increase the likelihood of drug repurposing. An example of this paradigm's effectiveness in the field of infectious illnesses is the approval of nifurtimox eflornithine combination therapy for the treatment of second-stage African trypanosomiasis [52]. The following sections will explain how precision and system medicine, in their own unique ways, present a new opportunity to expand the use of drug repurposing.

Lastly, displays with goals may end up being more beneficial. It is necessary to investigate the full range of compounds from a pharmaceutical business that have the same active scaffold once an approved medicine has proven effective against an undisclosed target. Hundreds of these compounds are typically manufactured and used in hit-to-lead and lead-optimization processes to define structure-activity connections; a lead compound used for one therapeutic goal may not be the same for another condition.

9.5 Predicting drug-target interactions during medication repurposing *via* multiple targeting

Comprehending the various binding behaviors of medications and developing novel theories for medication repurposing require precise modeling of the interactions between pharmaceuticals and their target proteins. Computational methodologies for forecasting drug-target interactions have garnered significant attention, primarily due to the challenges and expenses associated with experimental identification of binding interactions. It has been determined that there are three main types of computational approaches: target-based, ligand-based, and machine learning-based methods [53].

In ligand-based techniques, the putative ligand is evaluated in connection with established active ingredients that are directed toward the putative protein. This comparison helps forecast binding interactions since ligand-based methods, including pharmacophore modeling and Quantitative Structure–Activity Relationship (QSAR), depend on the presence of active ligands linked to the protein target in order to be effective [54].

Target-based techniques, which include methods such as binding-site similarity and docking, are effective means of distinguishing between protein-ligand interactions according to the target's three-dimensional structure. The limited availability of target structures, especially for G protein-coupled receptors (GPCRs), results in limitations [48, 54].

Novel drug-target pairings can be predicted by machine learning approaches by utilizing commonalities between compounds and targets. These methods can be generally divided into two categories: machine learning based on features and learning based on similarity. Additional subcategories of the similarity-based category include kernel-based, matrix factorization-based, and network-based approaches [49]. Machine learning techniques have the advantage of being quicker and more effective than docking procedures and data-intensive QSAR [50]. Nevertheless, popular databases have significant shortcomings, mainly because they include only true-positive interactions and leave out important information about the complex nature of drug–target interactions, like dose dependency and empirical affinities [51].

10. Conclusion

In summary, molecular docking is now a valuable tool in the drug repurposing market, offering a robust approach to discovering novel therapeutic applications for pharmaceuticals that have already received approval. Molecular docking is a promising avenue for drug development that could become more innovative and cost-effective in the future as technology advances and our understanding of molecular interactions deepens. In order to ensure that the potential of molecular docking for medicine repurposing is realized in an ethical and responsible manner, researchers need to keep improving and validating their techniques.

Author details

Ajay Kumar Shukla[1] and Joohee Pradhan[2]*

1 Institute of Pharmacy, Dr. Rammanohar Lohia Avadh University Ayodhya, India

2 Department of Pharmaceutical Sciences, Mohanlal Sukhadia University Udaipur, Rajasthan, India

*Address all correspondence to: juhipradhan123@gmail.com

References

[1] Naylor S, Kauppi DM, Schonfeld JP. Therapeutic drug repurposing, repositioning and rescue part II: Business review. Drug Discovery World. 2015;**16**(2):57-72

[2] Langedijk J et al. Drug repositioning and repurposing: Terminology and definitions in literature. Drug Discovery Today. 2015;**20**:1027-1034

[3] Wermuth CG. Selective optimization of side activities: The SOSA approach. Drug Discovery Today. 2006;**11**:160-164

[4] Begley CG, Ashton M, Baell J, Bettess M, Brown MP, Carter B, et al. Drug repurposing: Misconceptions, challenges, and opportunities for academic researchers. Science Translational Medicine. 2021;**13**:1-13

[5] March-Vila E et al. On the integration of in silico drug design methods for drug repurposing. Frontiers in Pharmacology. 2017;**8**:298-298

[6] Dovrolis N et al. Laying in silico pipelines for drug repositioning: A paradigm in ensemble analysis for neurodegenerative diseases. Drug Discovery Today. 2017;**22**:805-813

[7] Iwata H et al. Systematic drug repositioning for a wide range of diseases with integrative analyses of phenotypic and molecular data. Journal of Chemical Information and Modeling. 2015;**55**:446-459

[8] Nosengo N. Can you teach old drugs new tricks? Nature. 2016;**534**:314-316

[9] Naylor S, Schonfeld J. Therapeutic drug repurposing: Repositioning and rescue. Drug Discovery World. 2014;**16**:50-63. Available from: https://www.ddw-online.com/drug-discovery/p274232-therapeutic-drug-repurposingrepositioning-and-rescue-part-i:-overview.html [Accessed: February 12, 2020]

[10] Thayer A. Drug repurposing. Chemical & Engineering News. 2012;**90**(40). Available from: https://cen.acs.org/articles/90/i40/Drug-Repurposing.html [Accessed: February 12, 2020]

[11] Cavalla D. Scientific commercial value of drug repurposing. In: Dudley J, Berliocchi LE, editors. Drug Repositioning – Approaches and Applications for Neurotherapeutics. Abingdon: Taylor & Francis Group; 2017. pp. 3-22

[12] Hughes JP, Rees S, Kalindjian SB, Philpott KL. Principles of early drug discovery. British Journal of Pharmacology. 2011;**162**(6):1239-1249

[13] Rudrapal M, Khairnar SJ, Jadhav AG. Drug repurposing (DR): An emerging approach in drug discovery. Available from: https://www.intechopen.com/chapters/72744. DOI: 10.5772/intechopen.93193

[14] Rudrapal M, Chetia D. Endoperoxide antimalarials: Development, structural diversity and pharmacodynamic aspects with reference to 1,2,4-trioxane-based structural scaffold. Drug Design, Development and Therapy. 2016;**10**:3575-3590

[15] Kalita J, Chetia D, Rudrapal M. Design, synthesis, antimalarial activity and docking study of 7-chloro-4-(2-(substituted benzylidene)hydrazineyl) quinolines. Medicinal Chemistry. 2020;**16**(7):928-937. DOI: 10.2174/1573406415666190806154722

[16] Ferreira LG, Andricopulo AD. Drug repositioning approaches to parasitic diseases: A medicinal chemistry perspective. Drug Discovery Today. 2016;**21**(10):1699-1710

[17] Agrawal P. Artificial intelligence in drug discovery and development. Journal of Pharmacovigilance. 2018;**6**:2

[18] Cha Y, Erez T, Reynolds IJ, Kumar D, Ross J, Koytiger G, et al. Drug repurposing from the perspective of pharmaceutical companies. British Journal of Pharmacology. 2018;**175**(2):168-180

[19] Aggarwal S, Verma SS, Aggarwal S, Gupta SC. Drug repurposing for breast cancer therapy: Old weapon for new battle. Seminars in Cancer Biology. 2020;**68**:8-20. DOI: 10.1016/j.semcancer.2019.09.012

[20] Allarakhia M. Open-source approaches for the repurposing of existing or failed candidate drugs: Learning from and applying the lessons across diseases. Drug Design, Development and Therapy. 2013;**7**:753-766

[21] Jin G, Wong STC. Towards better drug repositioning: Prioritizing and integrating existing methods into efficient pipelines. Drug Discovery Today. 2014;**19**(5):637-644

[22] Ashburn TT, Thor KB. Drug repositioning: Identifying and developing new uses for existing drugs. Nature Reviews. Drug Discovery. 2014;**3**:673-683

[23] Oprea TI, Overington JP. Computational and practical aspects of drug repositioning. Assay and Drug Development Technologies. 2015;**13**:299-306

[24] Lionta E, Spyrou G, Vassilatis DK, Cournia Z. Structure-based virtual screening for drug discovery: Principles, applications and recent advances. Current Topics in Medicinal Chemistry. 2014;**14**:1923-1938

[25] Talevi A. Drug repositioning: Current approaches and their implications in the precision medicine era. Expert Review of Precision Medicine and Drug Development. 2018;**3**(1):49-61

[26] Kim TW. Drug repositioning approaches for the discovery of new therapeutics for Alzheimer's disease. Expert Review of Neurotherapeutics. 2015;**12**:132-142

[27] Rosa SGV, Santos WC. Clinical trials on drug repositioning for COVID-19 treatment. Revista Panamericana de Salud Pública. 2020;**44**:e40

[28] Turanli B, Grøtli M, Boren J, Nielsen J, Uhlen M, Arga KY, et al. Drug repositioning for effective prostate cancer treatment. Frontiers in Physiology. 2018;**9**:500

[29] Koch U, Hamacher M, Nussbaumer P. Cheminformatics at the interface of medicinal chemistry and proteomics. Biochimica et Biophysica Acta. 1844;**2014**:156-161

[30] Napolitano F, Zhao Y, Moreira VM, et al. Drug repositioning: A machine-learning approach through data integration. Journal of Cheminformatics. 2013;**5**(1):30

[31] Chong CR, Chen X, Shi L, Liu JO, Sullivan DJ Jr. A clinical drug library screen identifies astemizole as an antimalarial agent. Nature Chemical Biology. 2006;**2**:415-416

[32] Appleby BS, Cummings JL. Discovering new treatments for Alzheimer's disease by repurposing approved medications. Current Topics in Medicinal Chemistry. 2013;**13**:2306-2327

[33] Shah B, Modi P, Sagar SR. In silico studies on therapeutic agents for COVID-19: Drug repurposing approach. Life Sciences. 2020;**20**:117652

[34] Salentin S et al. From malaria to cancer: Computational drug repositioning of amodiaquine using PLIP interaction patterns. Scientific Reports. 2017;**7**:11401

[35] Pushpakom S, Iorio F, Eyers PA, et al. Drug repurposing: Progress, challenges and recommendations. Nature Reviews. Drug Discovery. 2019;**18**:41-58

[36] Pollastri MP. Fexinidazole: A new drug for african sleeping sickness on the horizon. Trends in Parasitology. 2018;**34**:178-179

[37] Garrido P, Aldaz A, Vera R, et al. Proposal for the creation of a national strategy for precision medicine in cancer: A position statement of SEOM, SEAP, and SEFH. Clinical & Translational Oncology. 2017;**41**(6):688-691

[38] Boolell M, Allen MJ, Ballard SA, Gepi-Attee S, Muirhead GJ, Naylor AM, et al. Sildenafil: An orally active type 5 cyclic GMP-specific phosphodiesterase inhibitor for the treatment of penile erectile dysfunction. International Journal of Impotence Research. 1996;**8**:47-52

[39] McBride WG. Thalidomide embryopathy. Teratology. 1977;**16**:79-82

[40] D'Amato RJ, Loughnan MS, Flynn E, Folkman J. Thalidomide is an inhibitor of angiogenesis. Proceedings of the National Academy of Sciences of the United States of America. 1994;**91**:4082-4085

[41] Singhal S, Mehta J, Desikan R, Ayers D, Roberson P, Eddlemon P, et al. Antitumor activity of thalidomide in refractory multiple myeloma. The New England Journal of Medicine. 1999;**341**:1565-1571

[42] Ning YM, Gulley JL, Arlen PM, Woo S, Steinberg SM, Wright JJ, et al. Phase II trial of bevacizumab, thalidomide, docetaxel, and prednisone in patients with metastatic castration-resistant prostate cancer. Journal of Clinical Oncology. 2010;**28**:2070-2076

[43] Rastegar-Mojarad M et al. Opportunities for drug repositioning from phenome-wide association studies. Nature Biotechnology. 2015;**33**:342-345

[44] Xu K, Cote TR. Database identifies FDA-approved drugs with potential to be repurposed for treatment of orphan diseases. Briefings in Bioinformatics. 2011;**12**:341-345

[45] Witkowski TX. Intellectual property and other legal aspects of drug repurposing. Drug Discovery Today: Therapeutic Strategies. 2011;**8**(3-4):139-143

[46] Breckenridge A, Jacob R. Overcoming the legal and regulatory barriers to drug repurposing. Nature Reviews. Drug Discovery. 2019;**18**:1-2

[47] Ritchie MD, Holzinger ER, Li R, et al. Methods of integrating data to uncover genotype-phenotype interactions. Nature Reviews. Genetics. 2015;**16**:85-97

[48] Hammond DJ, Cover B, Gutteridge WW. A novel series of chemical structures active in vitro against the trypomastigote form of Trypanosoma cruzi. Transactions of the Royal Society of Tropical Medicine and Hygiene. 1984;**78**:91-95

[49] Hammond DJ, Hogg J, Gutteridge WE. Trypanosoma cruzi: Possible control of parasite transmission

by blood transfusion using amphiphilic cationic drugs. Experimental Parasitology. 1985;**60**:32-34

[50] Bellera CL, Balcazar DE, Vanrell MC, et al. Computer-guided drug repurposing: Identification of trypanocidal activity of clofazimine, benidipine and saquinavir. European Journal of Medicinal Chemistry. 2015;**93**:338-348

[51] Alberca LN, Sbaraglini ML, Morales JF, et al. Cascade ligand- and structure-based virtual screening to identify new trypanocidal compounds inhibiting putrescine uptake. Frontiers in Cellular and Infection Microbiology. 2018;**8**:173

[52] Alirol E, Schrumpf D, Amici Heradi J, et al. Nifurtimox-eflornithine combination therapy for second-stage gambiense human African trypanosomiasis: Médecins Sans Frontières experience in the democratic republic of the Congo. Clinical Infectious Diseases. 2013;**56**(2):195-203

[53] Planer JD, Hulverson MA, Arif JA, et al. Synergy testing of FDA-approved drugs identifies potent drug combinations against Trypanosoma cruzi. PLoS Neglected Tropical Diseases. 2014;**8**(7):e2977

[54] Sykes ML, Avery VM. Development and application of a sensitive, phenotypic, high-throughput image-based assay to identify compound activity against Trypanosoma cruzi amastigotes. International Journal for Parasitology: Drugs and Drug Resistance. 2015;**5**(3):215-228

Chapter 9

Non-Coding RNAs Function as Diagnostic Biomarkers and Therapeutic Targets in Pulmonary Arterial Hypertension

Yingqi Wang, Daling Zhu and Hang Yu

Abstract

Pulmonary arterial hypertension is a disease characterized by complex pathogenesis and high mortality rates following diagnosis. Non-coding RNA plays a pivotal role in the development of pulmonary arterial hypertension, offering promising prospects as a diagnostic and therapeutic target for this condition. The utilization of nucleic acid drugs in disease treatment suggests the feasibility of packaging non-coding RNA into carrier systems and employing them in human pulmonary arterial hypertension (PAH) treatment through appropriate delivery routes. However, currently, no nucleic acid drugs are available for the clinical treatment of PAH. Identifying active regions within non-coding RNA through molecular docking analysis and developing suitable nucleic acid drugs hold great potential for advancing the field of PAH therapeutics.

Keywords: non-coding RNA, delivery, molecular docking, pulmonary arterial hypertension, clinical application

1. Introduction

1.1 Non-coding RNAs and pulmonary arterial hypertension

Pulmonary arterial hypertension (PAH) is a grave ailment characterized by the elevation in pulmonary artery pressure (mean pulmonary artery pressure $\geq$ 25 mmHg at rest) [1, 2], which results from the changes in the structure and function of the vessel wall due to aberrant proliferation, apoptosis, and migration of cells [3]. The causes of PAH can be broadly classified into primary and secondary causes. Currently, the pathogenesis of PAH is known to involve ion channels, vasoactive substances, immune factors, and genetic factors [4–6].

Recently, a growing body of evidence has highlighted the crucial regulatory roles played by non-coding RNAs (ncRNAs) in PAH development. It is worth noting that the majority of human genes (>95%) produce ncRNAs. MicroRNAs (miRNAs), which consist of 21–22 nucleotides, exert post-transcriptional control over gene expression [7]. Long non-coding RNAs (lncRNAs) are transcribed from intergenic or intragenic regions and typically exceed 200 nucleotides in length (**Figure 1**). They possess the ability to interact with proteins, RNA molecules, or DNA sequences to fulfill their regulatory functions [9]. Circular RNAs (circRNAs) represent a novel class of ncRNAs characterized by their closed-loop structure, which confers exceptional stability and enables interactions with both proteins and RNA molecules [10].

NcRNAs have been identified as regulators of multiple stages in gene expression. However, due to their vast quantity and diverse mechanisms, comprehensively understanding the roles of ncRNAs is challenging. NcRNAs often exhibit cell specificity in expression level and function, which are used for various disease therapy including PAH. A successful ncRNA-based strategy is based on a comprehensive understanding of the regulatory relationships involving ncRNAs [11].

1.2 Patential applications of non-coding RNAs

1.2.1 As therapeutic targets

With ever-increasing advancements in RNA structure determination methods and the ever-expanding discoveries of novel RNA structures and cellular functions, RNA-based therapeutics are emerging as promising approaches for treating human diseases. In general, RNA-based therapeutics are categorized into two types: (1) therapeutic RNAs: RNAs target RNA, DNA, or protein to modulate their function, such as RNA aptamers, antisense oligonucleotides (ASO), small interfering RNAs (siRNA), and guide RNAs (gRNA). Therapeutic RNAs are currently undergoing active development in the field of gene therapy [12–15]. (2) Targeted RNAs: RNAs serve as a target for binding with drug-like small molecules [15, 16], which bears resemblance to drug discovery targeting proteins.

However, only approximately 1.5% of the human genome is responsible for protein coding [17–21]. Among these protein-encoding genes, only a small fraction

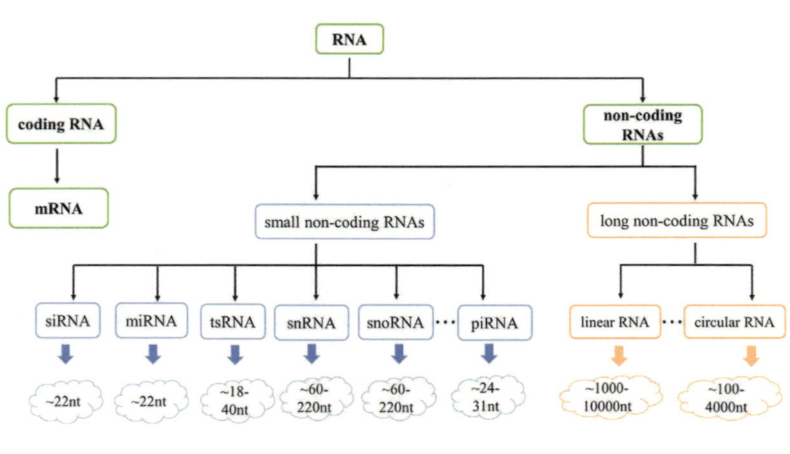

Figure 1.
RNA classification [8]. According to the size of non-coding RNAs (ncRNAs), they can be divided into >200 nucleotides (nt) long non-coding RNAs (lncRNAs) and small non-coding RNAs (sncRNAs) which are 18–200 nt. lncRNAs mainly include linear RNA and circular RNA. sncRNAs include small interfering RNA (siRNA), microRNA (miRNA), tRNA-derived small RNA (tsRNA), small nuclear RNA (snRNA), small nucleolar RNA (snoRNA), PIWI-interacting RNA (piRNA) and so on.

of about 10–15% are associated with diseases [18–20, 22–24]. Therefore, RNAs exhibit a much broader druggability than proteins. NcRNAs play crucial roles in various human diseases. Targeting ncRNAs will provide huge opportunities for drug discovery.

Gene therapy is a complex approach to treating diseases, which was first proposed by Theodore Friedmann and Richard Roblin in 1972 in "Science". RNAs, as a new class of treatments, may be an alternative to DNA therapy in preventing diseases. RNAs function in the cellular cytosol, avoiding the potential risks of insertion into patient genomes. RNAs have demonstrated therapeutic potential primarily through siRNA and microRNA (miRNA) applications, as evidenced by previous research efforts [25, 26]. In recent developments, mRNA vaccines have emerged as a powerful tool in combating infectious diseases such as COVID-19 [27–29].

NcRNAs have the potential to be PAH therapeutic targets. MiR-140-5p was found to be downregulated in PAH model rats; miR-140-5p mimics prevented the progression of established PAH in rats and suppressed the proliferation and migration of pulmonary artery smooth muscle cells (PASMCs) [30]. The results are promising. However, the clinical application of ncRNA therapy is currently constrained by several factors, including the diverse biological functions of ncRNAs and challenges in formulation and delivery approaches. As a therapeutic agent, ncRNAs may elicit adverse effects that could pose life-threatening risks or result in inadequate deposition within the deep lung tissues. Furthermore, large-scale, comprehensive, and well-designed clinical studies are indispensable for shedding more light on the potential applications of ncRNAs.

1.2.2 As biomarkers

NcRNAs have emerged as potential non-invasive and more sensitive diagnostic biomarkers for PAH. In the early diagnosis of PAH, hsa-miR-21-3p and hsa-miR-143-3p in peripheral plasma can be considered as potential biomarkers [31].

1.3 Status of pulmonary non-coding RNAs delivery

RNA therapeutics have been hailed as the medicine of the future, and this vision has now become our reality [32]. Currently, there are two mRNA vaccines available in the market, along with four siRNA-based therapeutics and five ASO-based drugs [33]. This signifies a true revolution in the field of RNA therapeutics; however, one major obstacle remains delivery [34]. Nanocarriers that are optimized for protecting and delivering payloads of RNA tend to accumulate primarily in the liver when administered intravenously [35]. To overcome this limitation and explore alternative administration routes beyond liver targeting, numerous clinical trials have investigated different approaches for RNA delivery [36]. Intravitreal injection (fomivirsen) and intrathecal injection (nusinersen) have demonstrated success, prompting further exploration of local administration routes such as intranasal or pulmonary delivery through ongoing clinical trials. However, no approved products utilizing these routes have emerged thus far [37]. The lung presents an array of currently untargetable sites that hold potential for treatment using ncRNA therapeutics.

In this chapter, we conducted a comprehensive literature review to identify functional ncRNAs, particularly miRNAs, circRNAs, and lncRNAs associated with PAH. Additionally, we identified potential untapped biomarkers and therapeutic targets for ncRNAs in PAH.

2. MicroRNAs

2.1 Introduction

MicroRNAs (miRNAs) are a class of ncRNAs, approximately 20–24 nucleotides in length, that possess regulatory functions. MiRNA genes are transcribed by RNA polymerase II or III into primary miRNA transcripts (pri-miRNA). These pri-miRNAs are subsequently cleaved by the Drosha-DGCR8 (DiGeorge syndrome critical region 8) complex [38], an RNase III enzyme. The resulting precursor miRNAs (pre-miRNAs) are then transported to the cytoplasm and processed into double-stranded mature miRNAs by the RNase Dicer [39]. This double-stranded form is rapidly loaded onto the RNA-induced silencing complex (RISC), where one of the single strands undergoes degradation while the other mature single-stranded miRNA molecule either degrades or translates its target gene through complementary pairing with the 3′UTR region of said target gene [40].

The stability of miRNAs is remarkably high in both DNA and RNA lysates, making them resistant to degradation. MiRNAs exhibit conserved sequences and functions across diverse cells, tissues, organs, and genera. The distinct expression profiles of miRNAs in various tissues and cells render them valuable molecular markers for specific cellular populations. Importantly, the presence of a particular miRNA at a specific stage within a given cell determines its differentiation direction and phase. Consequently, miRNAs play crucial roles as pivotal regulators governing cell timing and directional differentiation.

Numerous studies have demonstrated that miRNAs exert their biological function by participating in the regulation of downstream gene translation processes, with target mRNAs located within its 3′UTR [41]. In animals, a single miRNA can recognize multiple mRNA targets, and conversely, one mRNA target can be recognized by multiple miRNAs. It is estimated that approximately two-thirds of protein-coding genes in the human genome are regulated by miRNAs.

2.2 MiRNAs and PAH

It is widely acknowledged that miRNA dysfunction contributes to the enhanced proliferation of endothelial cells, smooth muscle cells, and outer epidermal cells in PAH. Increasing evidence suggests that various miRNAs and their expression changes (either upregulation or downregulation) are associated with PAH pathogenesis. MiRNAs regulate the proliferation and apoptosis of pulmonary artery endothelial cells (PAECs), PASMCs, and extrapulmonary fibroblasts through their target genes. They play roles in all aspects of pulmonary vasoconstriction and reconstruction. However, due to a large number of publications based on diverse experimental methods, it remains challenging to identify the most crucial miRNAs as potential therapeutic targets for PAH [42].

The level of miR-130/301 has been shown to be elevated in both pulmonary vessels and plasma in mammalian models as well as human patients with PAH. MiR-130/301 promotes PAH-related phenotypes through the apelin-miR-424/503-FGF2 signaling pathway in endothelial cells, while it utilizes the STAT3/miR-204 signaling pathway in smooth muscle cells. Administration of miR-130a oligonucleotide mimics enhances pathogenic effects related to PAH, whereas inhibition of miR-130/301 prevents the onset of PAH in mice treated with SU5416 under hypoxic conditions [43].

The miRNA-17/92 cluster exerts an impact on the proliferation of PAECs and PASMCs. The combination of miR-17/18a/19a/92a demonstrates the most potent induction effect on PCNA and α-SMA, while maintaining calponin and SM22α levels. Consequently, in SM-17 ~ $92^{-/-}$ mice, miR-17 to 92 was reconstituted by administering a mixture of miR-17/18A/19a/92a mimics via intravenous injection through the tail vein. MiR-17 to 92 reconstitution restored elevated right ventricular systolic blood pressure (RVSP), RV/(LV + S) ratios, and pulmonary artery remodeling in hypoxia-exposed SM-17 ~ $92^{-/-}$ mice without altering basal levels of RVSP, RV/(LV + S), or pulmonary artery wall thickness. These findings suggest that miR-17 ~ 92 plays a crucial role in the pathogenesis of PAH. By investigating PASMCs and lung tissue from PAH patients, it has been demonstrated that miRNA-17/92 promotes PASMC proliferation through the PDLIM5/TGF-beta/Smad pathway, offering novel insights for PAH treatment [44].

The expression of miR-495 is significantly upregulated in the mouse model of PAH induced by chronic hypoxia and hypoxia-Su5416 (HySu). Inhibition of miR-495 via adeno-associated virus type 9 (AAV9) effectively mitigates hemodynamic abnormalities and structural changes in the pulmonary vasculature in the mouse model of PAH, thereby improving vascular remodeling and angiogenesis. This represents a novel potential therapeutic approach for human pulmonary hypertension [45].

The serum level of miR-30a was found to be elevated in patients with pulmonary hypertension compared to the healthy control group. MiR-30a exhibits predominant expression in the pulmonary arteriole (PA). In animal models of HySu-induced and MCT-induced PAH, genetic knockout of the miR-30a gene effectively reduces right ventricular systolic blood pressure (RVSP), as well as attenuates remodeling of the pulmonary arteriole and hypertrophy of the right ventricle. Furthermore, pharmacological inhibition of miR-30a through intratracheal fluid perfusion (IT-L) administration strategy demonstrates high efficacy [46].

Our previous studies have demonstrated that circ-calm4 promotes pyroptosis in hypoxic PASMCs through the miR-124-3p/PDCD6 pathway [47]. Additionally, this circular RNA also enhances PASMC proliferation via the miR-337-3p/Myo10 pathway [48]. These findings suggest that circ-calm4 may represent a promising therapeutic target for treating pulmonary hypertension.

2.3 MiRNAs in clinical research

In terms of clinical diagnosis, accessing certain tissues such as the cardiac tissue can be challenging. However, miRNAs have emerged as potential biomarkers for disease states and prognostic markers in various diseases. An analysis conducted on plasma samples from 64 untreated patients with pulmonary arterial hypertension, along with 43 patients and healthy controls, demonstrated that miR-636 and miR-187-5p exhibit high diagnostic accuracy in predicting pulmonary hypertension [49]. The findings of these studies indicate that miRNAs have the potential to serve as biomarkers for the diagnosis of PAH.

A significant number of clinical trials involving ncRNAs primarily focus on miRNAs, with over 800 trials conducted, a substantial proportion of which are specifically targeted toward cancer research [50]. Growing evidence suggests that miRNAs play crucial roles in clinical diagnosis and therapy (**Figure 2**). The first miRNA to transition from basic research to clinical application is miR-122, which has been utilized for the treatment of hepatitis C virus (HCV) [52]. MRX34 represents the pioneering miRNA-based approach for cancer treatment, where it is encapsulated

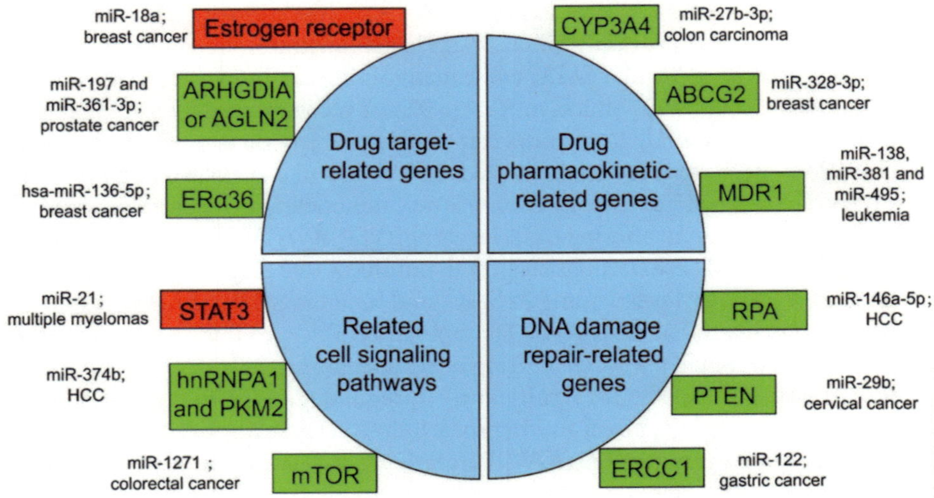

Figure 2.
Several miRNA-related cancer therapy resistance [51]. There are four categories along with some regulators presented in the text, where the rectangle in green or in red shows whether the function of miRNAs in the mentioned cancer type is positive or negative for the corresponding therapy. ERα36: estrogen receptor alpha 36; CYP: cytochrome P450; ABCG2: ATP-binding cassette sub-family G member 2; MDR1: multidrug resistance 1; STAT3: signal transducer and transcriptional activating factor 3; MYC: myelocytomatosis oncogene; HnRNPA1: heterogeneous nuclear ribonucleoprotein A1; PKM2: the pyruvate kinase muscle 2; mTOR: mammalian target of rapamycin; RPA: replication protein A; PTEN: phosphatase and tensin homologous protein; ERCC1: excision repair cross-complementing 1.

within lipid nanoparticles and administered via intravenous injection into the human body [53]. Furthermore, a nanoparticle formulation named TTX-MC138 has been developed to target anti-miR-10b for breast cancer treatment [54], while an ointment incorporating ionic liquid technology and a mimic of ds-miR-634 has been formulated for skin squamous cell carcinoma (cSCC) therapy [55].

The dysregulation of miRNA expression in exosomes has been demonstrated to be a potential therapeutic target for PAH, with the upregulation of miR-21 observed compared to the matched control group [56]. However, there is currently a lack of clinical studies on the therapeutic application of miRNAs in pulmonary hypertension. The future holds immense potential for nucleic acid therapy of pulmonary hypertension, as the combination of meticulous selection of miRNAs from fundamental research and efficient delivery routes will pave the way for groundbreaking advancements in this field.

3. Circular RNAs

3.1 Introduction

Circular RNAs (circRNAs) are a newly discovered class of ncRNAs that lack a 5′ terminal cap and a 3′ terminal poly(A) tail, instead forming a circular structure through covalent bonds [57, 58]. This unique circular conformation distinguishes circRNAs from conventional linear RNAs.

The categorization of circRNAs is based on their origin, which can be classified into three categories: exonic circRNAs derived from exons, circular intronic RNAs derived from introns, and retained-intron circRNAs composed of both exons and introns [59].

CircRNAs possess diverse biological capabilities, including the regulation of target gene expression and modulation of target gene activity through competitive binding with corresponding miRNAs, thereby influencing disease occurrence and progression. These circRNAs are commonly referred to as competitive endogenous RNAs (ceRNAs), acting as molecular "sponges" [60]. Furthermore, circRNAs interact with RNA-binding proteins [61], regulate mRNA stability [62], and serve as protein molecule sponges, with some circRNAs containing binding sites for one or more RNA-binding proteins [63]. Despite being classified as ncRNAs, a subset of circRNAs can encode peptides that exert regulatory functions through peptide encoding [64]. The findings of circRNAs pave the way for future investigations into circRNA biology, offering a novel perspective on our comprehension of gene expression regulation and disease pathogenesis.

3.2 CircRNAs and PAH

CircRNAs exhibit aberrant expression patterns in various diseases, including cancer, neurological disorders, and immune diseases [65–68]. Additionally, they play a pivotal role in the pathogenesis of pulmonary hypertension.

CircRNAs exhibit aberrant expression in patients with PAH compared to healthy controls, and are involved in the processes of cell proliferation, angiogenesis, inflammation, and fibrosis in PAH patients. This suggests their potential role in disease progression [69–71]. The circHIPK3/miR-328-3p/STAT3 axis contributes to the pathogenesis of PAH by promoting proliferation, migration, and angiogenesis of human PAECs. The accelerated involvement of circHIPK3 indicates its potential value as a therapeutic target for treating PAH [72]. Hsa_circNFXL1_009 directly acts as a sponge for hsa-miR-29b-2-5p (miR-29b), positively regulating the expression of voltage-gated potassium (K^+) channel subfamily B member 1 (KCNB1) at the mRNA level. Delivery of exogenous hsa_circNFXL1_009 attenuates hypoxia-induced proliferation, resistance to apoptosis, and migration of human PASMCs, suggesting its potential use as a therapeutic target for treating PAH [73].

3.3 CircRNAs in clinical research

CircRNAs have emerged as significant markers and targets for the diagnosis and treatment of numerous diseases owing to their distinctive structure, which effectively shields them from exonuclease degradation, rendering them relatively stable in terms of drug efficacy and biological activity (**Figure 3**).

The abundant presence and remarkable stability of circRNAs in bodily fluids render them an optimal non-invasive biomarker. The identification of circRNAs in serum or tissues holds the potential for early screening, disease monitoring, and prognosis evaluation.

It has been suggested that serum exosomal circCCDC66 secreted and released by pituitary adenoma cells can serve as a potential prognostic biomarker for monitoring tumor dynamics during treatment [76], which highlights the utility of circRNAs as biomarkers for early disease screening, disease monitoring and prognosis evaluation. Researchers have investigated the distinctive expression pattern of hsa_circ_0003416 in the plasma of pediatric patients with PAH resulting from congenital heart disease (CHD). It was observed that hsa_circ_0003416 exhibited a significant downregulation in children affected by PAH-CHD and demonstrated its potential as a biomarker for diagnosing PAH-CHD [77]. This finding implies that circRNAs could serve as valuable diagnostic markers for PAH in clinical settings.

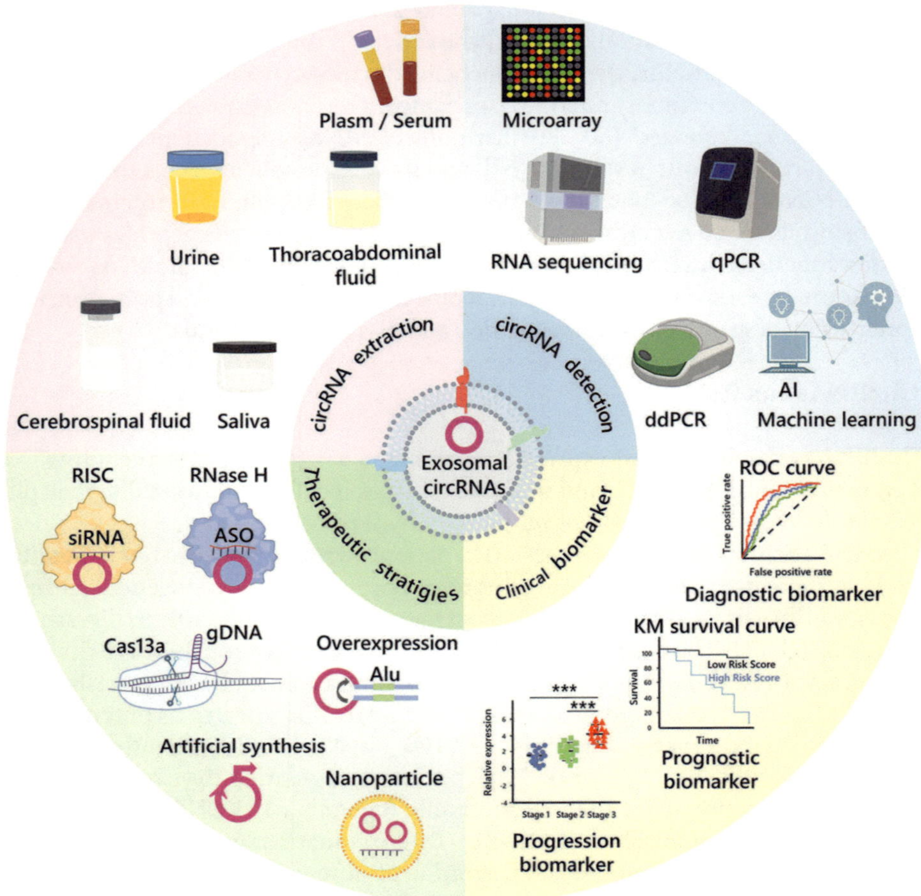

Figure 3.
Clinical applications of circRNAs and exosomal circRNAs in tumors, including circRNA extraction, detection, as clinical markers and therapeutic strategies [74, 75].

CircRNAs have been extensively investigated in various diseases. For instance, researchers have observed that Steatohepatitis-associated circRNA ATP5B Regulator (SCAR), localized in mitochondria, exerts inhibitory effects on mitochondrial ROS (mROS) production and fibroblast activation. To explore the therapeutic potential of circRNA SCAR *in vivo*, they intravenously administered mito-NP containing circRNA SCAR, resulting in a significant increase of its hepatic levels in mice and attenuation of insulin resistance induced by high-fat diet-induced cirrhosis [78]. Additionally, it has been demonstrated that circNR3C2 negatively regulates triple-negative breast cancer metastasis through the circNR3C2/miR-513a-3p/HRD1/Vimentin axis, and this finding was validated using mouse models. Notably, circNR3C2 exhibits inhibitory effects on TNBC metastasis [79]. Collectively, these studies highlight the promising therapeutic potential of targeting circRNAs for clinical applications.

Our previous study discovered that the circ-calm4/miR-337-3p/Myo10 signaling pathway regulates the proliferation of PASMCs at the molecular level and validated this finding in mice. Administration of serotype 5 adenovirus-associated virus (AAV5) through nasal feeding revealed that silencing circ-calm4 alleviated symptoms associated with pulmonary hypertension [48]. Lihuang Su et al. reported that circ-Ntrk2 acts as a sponge for miR-296-5p and miR-296-5p, which bind to the 3′-untranslated region of transforming growth factor beta 1 (TGF β1) mRNA, thereby attenuating TGF β1 translation. Additionally, knocking down circ-Ntrk2 in mice resulted in improved symptoms of PAH [80]. These studies suggest that circRNAs hold potential as therapeutic targets for treating pulmonary hypertension.

Despite the potential of circRNAs in clinical treatment, there are still several challenges that need to be addressed. Antisense technology can selectively inhibit or degrade oncogenic circRNAs, but most loss-of-function studies have only been conducted on preclinical animal models using RNAi. Currently, no circRNA-targeted treatments have reached clinical trials yet [81]. However, delivery of circRNA expression cassettes can be achieved through encapsulation within lipid and/or polymer nanoparticles that can even target specific organelles like mitochondria [82]. Synthetic approaches have also been developed for producing and delivering pre-formed circRNAs to target cells using non-viral nanoparticles [83]. Additionally, researchers have synthesized artificial circRNA sponges called "circmiRs" *in vitro* to therapeutically target miRNAs [84–87], while other studies show promising results in using circRNAs as vaccines for COVID-19 [88] and melanoma [89]. These findings provide valuable insights into the future application of circRNAs in treating PAH.

In summary, the current research on circRNA clinical application is still in its early stages. Its future prospects for development in scientific research and clinical investigations are extensive. Studies are required to validate its efficacy and reliability in clinical diagnosis. There is an urgent necessity to develop and enhance advanced methodologies for the design, synthesis, purification, delivery, and therapeutic applications of *in vitro* artificial circRNA vaccines or drugs. In the future, researchers need to strengthen collaboration, data acquisition, and resource utilization while promoting the translation of circRNA research findings into clinical practice to achieve improved treatment outcomes and enhanced quality of life for patients.

4. Long non-coding RNAs

4.1 Introduction

Long non-coding RNAs (lncRNAs) are characterized by their extensive nucleotide sequences, exceeding 200 nucleotides in length, making them a diverse and intriguing group of RNA molecules. LncRNAs play crucial roles in regulating various cellular processes [90].

The biogenesis of lncRNAs follows mechanisms similar to mRNA, involving transcription by RNA polymerase II, 5′ capping, splicing, and 3′ polyadenylation [91]. However, unlike mRNAs, lncRNAs rarely encode proteins and instead exert their functions through a myriad of intricate molecular mechanisms, including acting as molecular scaffolds [74, 75], guiding the assembly of protein complexes [92], serving as decoys by sequestering regulatory proteins [93], and functioning as enhancer-associated transcripts to modulate gene expression over long genomic distances [94].

One notable characteristic of lncRNAs is their specific expression patterns in different tissues and developmental stages [95]. These dynamic expression profiles reflect their involvement in essential biological processes, including embryonic development, cellular differentiation, and tissue homeostasis maintenance. The dysregulation of lncRNAs has increasingly been associated with various human diseases, highlighting their potential as diagnostic markers and therapeutic targets.

4.2 LncRNAs and PAH

LncRNAs play significant roles in cardiovascular disorders by influencing vascular function, cardiac development, and fibrosis [96–100]. An increasing body of evidence suggests that lncRNAs serve as crucial epigenetic modifiers in pulmonary hypertension. The dysregulation of lncRNAs may impact processes such as vascular remodeling [101], inflammation [102], and endothelial dysfunction [103], thereby contributing to the narrowing of pulmonary vessels observed in PAH [104].

A research from 11 control individuals and 12 pulmonary arterial hypertension patients suggests that JPX modulates RABEP1 (Rab GTPase-binding effector protein expression through miRNAs (miR-145, miR-146 ac, miR-216a, miR-216b, miR-24, miR-33ab, miR-129-5p), thereby impacting $CD8^+$ T cells, NK cells, eosinophils, and other immune cells, ultimately influencing the immune processes associated with PAH. The finding reveals that lncRNA-mRNA pair (JPX-RABEP1) can serve as a potential biomarker in PAH [105]. The rs619586A > G polymorphism in MALAT1 plays a protective role in susceptibility to PAH within the Chinese Han population. This polymorphism has the potential to serve as a novel biomarker for PAH susceptibility and implicates lncRNAs in the etiology of PAH [106]. A study from 41 blood samples, SNHG12 is identified as a competing endogenous RNA (ceRNA) of CCR7 by sequestering hsa-let-7e-5p, which suggests the potential of CCR7, hsa-let-7e-5p, and SNHG12 as candidate biomarkers for PAH.

Although lncRNAs have been reported as potential biomarkers in exosomes or body fluids in diseases such as cancer, they are still in the developmental stage in the field of pulmonary arterial hypertension, but these studies may shed light on the discovery of biomarkers in pulmonary hypertension. Identification of lncRNAs as potential biomarkers holds promise for disease diagnosis, prognosis, and monitoring. The development of non-invasive diagnostic tools based on lncRNA expression profiles could significantly impact early disease detection and patient management.

A study from our team proves that m6A-induced degradation of FENDRR promotes pyroptosis in human pulmonary artery endothelial cells (HPAECs) by regulating the methylation of the DRP1 promoter, providing a potential novel target for hypoxia-induced pulmonary hypertension (HPAH) therapy. The overexpression of the conserved fragment TFO2 of FENDRR prevents HPAH *in vivo* [107]. The lncRNA PAXIP1-AS1 facilitated the proliferation and migration of human PASMCs through the ETS1/WIPF1/RhoA pathway. Inhibition of PAXIP1-AS1 via tail vein injection of lentiviral plasmid expressing stable short hairpin RNAs (sh-RNAs) specifically targeting PAXIP1-AS1 significantly attenuated the elevated right ventricular systolic pressure (RVSP) in MCT-induced pulmonary hypertension mice. Therefore, targeting lncRNA PAXIP1-AS1 may hold promise as a potential therapeutic strategy for PAH treatment [108]. The expression of lncRNA H19 is significantly elevated in rodent lungs induced by MCT and is upregulated by PDGF-BB. The H19/let-7b/AT-1R axis contributes to the pathogenesis of PAH by promoting the proliferation of pulmonary artery smooth muscle cells. Knockout of H19 plays a protective role in the development of PAH, suggesting that targeting H19 may hold potential for PAH treatment [102].

4.3 LncRNAs in clinical research

RNAs have demonstrated therapeutic potential primarily through small interfering RNA (siRNA) and microRNA (miRNA) applications [25, 26]. LncRNAs have gained significant attention in clinical research due to their diverse roles in various physiological and pathological processes. Advances in understanding the roles of lncRNAs may pave the way for developing novel therapeutic strategies. Targeting disease-associated lncRNAs could offer new avenues for precision medicine, especially in the treatment of pulmonary hypertension.

Given that lncRNA falls in size between siRNA and mRNA, its delivery vehicle may not differ significantly from those already employed in clinical settings for siRNA and mRNA delivery [109]. Therefore, leveraging existing delivery platforms could facilitate the translation of lncRNA-based therapeutics into clinical practice. However, the intricate secondary structure of lncRNAs poses challenges for their exogenous delivery, hindering the development of lncRNA-based drugs. As a result, instead of directly transferring lncRNAs, many lncRNA-based drugs opt to target lncRNAs to modulate their expression. The next is an exploration of diverse delivery systems utilized for RNA therapeutics, with some having undergone testing for lncRNA and nucleic acid delivery targeting lncRNAs.

Numerous studies have explored the potential of lncRNAs as diagnostic and prognostic biomarkers in various cancers, such as LINC01535 in breast cancer [110], MLETA1 in lung cancer [111], and HCG18 in cholangiocarcinoma [112]. LncRNAs ANRIL and HOTAIR can be used as biomarkers for cardiovascular disease risk prediction [113]. LncRNA PCA3 is highly expressed in prostate cancer and is considered useful for prostate cancer diagnosis [112]. However, lncRNAs as PAH biomarkers are still being researched.

5. PIWI-interacting RNA

5.1 Introduction

PiRNAs (PIWI-interacting RNA) are a class of small non-coding RNA molecules, typically ranging from 24 to 32 nucleotides in length [114]. PiRNAs exhibit highly conserved and diverse sequences. Unlike other types of small RNAs such as miRNAs and siRNAs, piRNAs are primarily generated through the interaction between transposon factors and the genome [115]. PiRNAs are mainly expressed in the reproductive system. They form complexes with PIWI proteins to form piRNA-induced silencing complexes (piRISCs) that play crucial roles in protecting genome stability during transposition, suppressing transposon activity, and regulating gene expression and epigenetic modifications [116]. The initial belief that piRNAs are exclusively involved in germ cell development has been challenged by recent research, which has demonstrated their expression in various tissues and cell types. Furthermore, these findings highlight the involvement of piRNAs in maintaining immune homeostasis and underscore their significance in diseases such as cancer [117].

5.2 PiRNAs and PAH

The involvement of piRNAs in crucial processes such as cardiovascular remodeling, inflammatory response, and cell proliferation is closely associated with the pathophysiological mechanisms underlying pulmonary arterial hypertension. PiRNA-63076 is upregulated in hypoxic PASMCs and the lung tissues of hypoxia-induced mice, which promotes the proliferation of PASMCs by regulating the interaction of acyl-CoA dehydrogenase (Acadm) with Piwi protein. PiRNA-63076 partially inhibits the expression of Acadm by mediating DNA methylation in the promoter region of Acadm, providing a new perspective on understanding the pathophysiological process of pulmonary arterial hypertension [118]. Some piRNAs from the extracellular vesicles of PAH patients exhibited different expression patterns correlated with the severity of PAH, indicating a potential role for piRNAs in the development and progression of PAH [119].

5.3 PiRNAs in clinical research

The expression of piRNAs demonstrates disease-specificity, rendering them potential diagnostic biomarkers. Researchers have already employed piRNAs for early tumor detection and prediction of patient prognosis, showcasing promising clinical applicability [120]. Assessing piRNAs levels in blood, urine, or other bodily fluids may be feasible to aid in the early diagnosis, classification, and prognostic evaluation of diseases.

PiRNAs also play a role in the treatment of diseases. PiRNA-54265 is upregulated in colorectal cancer. AntagopiR54265, the inhibitor of piRNA-54265, significantly suppressed the growth and metastasis of implanted tumors [121]. By targeting piRNAs as therapeutic targets, the regulation of relevant gene expression can be achieved through interference with specific interactions between piRNAs and Piwi proteins. Through the design and synthesis of specific piRNA ligands, it is possible to precisely modulate gene expression, signal pathways, and cellular processes, thereby intervening in disease development.

The research on PiRNAs is still in its nascent stage. The clinical application of piRNAs is still in the research stages, requiring further experiments and clinical studies to validate its accuracy, feasibility, and safety. PiRNAs hold promise as diagnostic biomarkers for various diseases, a powerful tool for targeted therapy, and a strategic approach toward cell therapy and regenerative medicine [120].

6. Small nucleolar RNA

6.1 Introduction

SnoRNAs (small nucleolar RNA) are a class of small non-coding RNA molecules that mainly exist in the nucleolus. They play regulatory roles in the nucleolus and are involved in the modification and maturation processes of ribosomal RNAs (rRNAs) [122]. SnoRNAs are typically composed of 60–300 nucleotides and are divided into two main categories: C/D box snoRNAs and H/ACA box snoRNAs. C/D box snoRNAs mainly form nucleotide methyltransferase complexes with proteins to carry out 2′-O-methylation modifications at specific positions in rRNAs. H/ACA box snoRNAs form complexes with proteins to maintain RNA stability and localization and guide post-transcriptional modifications in rRNAs [123].

SnoRNAs are involved in various cellular processes, such as proliferation, transcription, and RNA splicing, which are associated with tumor development and metastasis [124]. The abnormal expression or dysfunction of snoRNAs may play significant roles in the development of diseases, including cardiovascular diseases, congenital disorders, and genetic diseases [125].

6.2 SnoRNAs and PAH

A study from different types of pulmonary arterial hypertension patients (including idiopathic pulmonary arterial hypertension (IPAH), connective tissue disease-associated pulmonary arterial hypertension (CTD-PAH), and congenital heart disease-associated pulmonary arterial hypertension (CHD-PAH)) shows that there are 30 differentially expressed snoRNAs in lung tissues [126]. SNORD20, SCARNA4, and SNORD94 exhibit consistent differential expression in all three PAH, suggesting the three snoRNAs play important roles in the pathophysiological processes of PAH [126].

The relationship between snoRNAs and diseases still encompasses numerous unexplored areas. Further investigations are imperative to accumulate additional evidence elucidating the association between snoRNAs and diseases, thereby offering novel insights and approaches for disease prevention and therapy.

7. tRNA-derived small RNA

7.1 Introduction

TsRNA (tRNA-derived small RNA) refers to a class of small RNAs generated from tRNAs. TRNAs are a common type of RNA that primarily transports amino acids to the ribosome during protein synthesis. However, recent research has discovered that tRNAs participate in other cellular processes through various mechanisms and give rise to various tsRNAs [127]. The generation of tsRNAs mainly involves two mechanisms: tRNAs cleavage and modification and the generation of cleavage products through non-canonical functions of tRNAs. TsRNAs have diverse biological functions, including regulation of gene expression, translational control, and impact on cellular processes such as growth, division, and apoptosis [128]. Abnormal expression of tsRNAs may be associated with the occurrence and development of tumors. For example, tsRNAs are significantly increased or decreased in cancer cells, which may affect the expression of tumor suppressor genes and promote the growth and spread of tumor cells [129].

7.2 tsRNAs and PAH

An increasing number of studies are investigating the association of tsRNAs with cardiovascular diseases, such as pulmonary hypertension [32, 97, 99]. Altered expression of tsRNAs in the blood and lung tissue of PAH patients has been found. A study conducted a small RNA microarray analysis of tsRNAs in the plasma of pulmonary arterial hypertension patients and a healthy control group identified a total of 816 tsRNAs significantly regulated in PAH, with 243 upregulated and 573 downregulated [130]. Gene Ontology analysis showed that these tsRNAs might be involved in cellular metabolic conditions and other processes, suggesting that tsRNAs could potentially serve as novel biomarkers and therapeutic targets in idiopathic PAH [130].

8. Delivery tools for RNAs

Gene therapy is extensively utilized for the treatment of incurable disorders and has become a routine procedure in clinical practice. The main challenge in the development and application of RNA drugs lies in their delivery technology, as opposed to small molecule and protein drugs. Unlike these counterparts, RNA molecules are negatively charged, sensitive to ubiquitous RNases, and primarily act intracellularly.

Therefore, the advancement of delivery systems is crucial for addressing existing issues *in vivo* RNA delivery and enhancing the efficacy of RNA drugs (**Figure 4**).

8.1 Viral vector for RNA delivery

Viral vectors have emerged as indispensable components of gene therapy due to their inherent ability to infect mammalian cells and effectively deliver genetic material [132]. Adeno-associated virus (AAV), adenovirus (AdVs), and lentiviral vectors (LVV) are increasingly being recognized as fundamental therapeutic delivery systems owing to their usability and safety profile [133].

However, the utilization of viral vectors can give rise to two types of adverse effects: a response triggered by the virus's entry or a reaction induced by the delivered

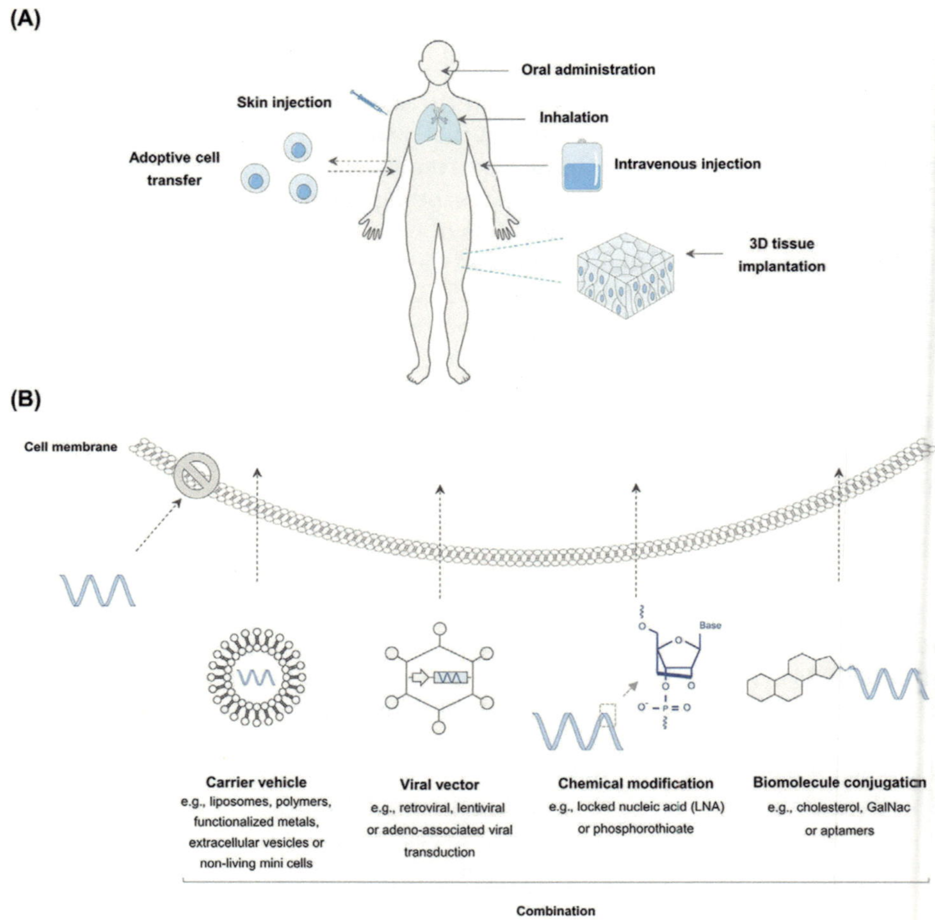

Figure 4.
Schematic overview of administration routes and features of miRNA therapeutics [131]. (A) miRNA therapeutics can be applied by systemic administration (e.g., by venous injection) or by cell–/tissue-directed administration, extracorporeal miRNA approaches, adoptive cell transfer, or implantation of 3D matrices releasing miRNA therapeutics. (B) Nucleic acid-based miRNA therapeutics show a low cell membrane permeability. Efficient intracellular delivery can be achieved by nano-sized carrier vehicles, viral transduction, introduction of chemical modifications, or conjugation to biomolecules that further a receptor-mediated uptake. Abbreviations: GalNAc, N-acetylgalactosamine; miRNA, microRNA.

genetic material. Any viral-based gene therapy approach necessitates thoroughly evaluating potential side effects and safety concerns.

AAV is a small, non-enveloped virus requiring helper viruses such as AdV, herpesvirus, and some others for replication. AAV is a natural human symbiont, and a weak immune response makes it a suitable vector for gene therapy [134–136].

AdV, a non-enveloped virus, contains protein-coding genes belonging to Adenoviridae [137, 138]. The wide applications of AdV are due to their high levels of gene expression, transduction efficiency, and ability to deliver transgenes of up to 8 kbp in size [139, 140].

Lentivirus is an enveloped virus belonging to the family Retroviridae. The advantages of LVVs are the packaging capacity of up to 10 kb [141], low cytotoxicity, and lack of viral replication after transduction.

8.2 Non-viral vectors for RNA delivery

The limitations of viral vector delivery systems such as immunogenicity and mutagenicity restrict their clinical application and prompt the development of non-viral gene carriers using different materials to achieve successful *in vivo* delivery of RNA (**Figure 5**).

8.2.1 Nanoparticles

8.2.1.1 Lipid nanoparticles (LNP)

LNPs stand as the sole FDA-approved carrier for macromolecular nucleic acid drugs, facilitating enhancements in clinical trials [143, 144]. Three LNP-based RNA drugs have been approved on the market, including Alnylam's Patisiran (ONPATTRO™), Pfizer's BNT162b2, and Moderna's mRNA-1273. Despite LNPs exhibiting lower transfection efficiency compared to lentiviral vectors, their advanced, safe, relatively convenient, and cost-effective construction process renders them practical alternatives (**Figure 6**) [145].

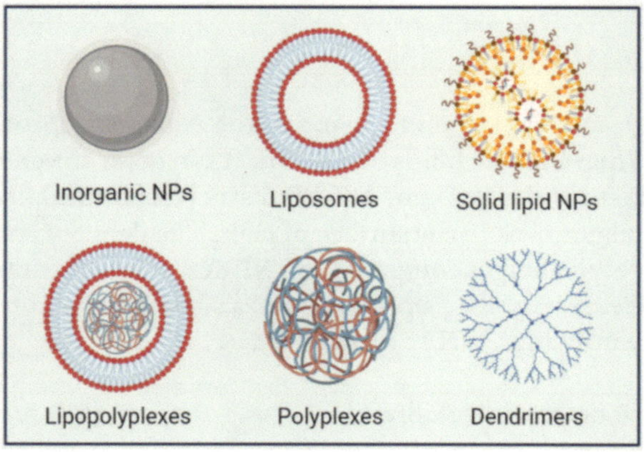

Figure 5.
A schematic representation of some nanocarriers used for miRNA delivery [142].

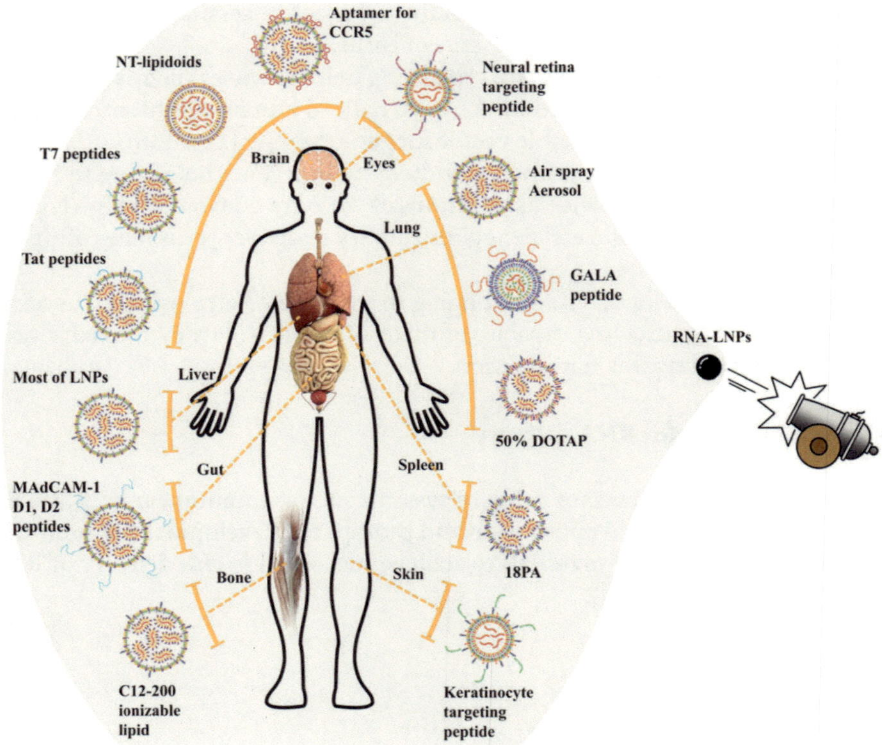

Figure 6.
Illustration of targeted therapy using sophisticated RNA-LNPs designed to target specific tissues and organs [97–99]. (1) LNPs expressing (a) Tat and T7 peptides, (b) NT-lipidoids, and (c) aptamer for CCR5 can be preferentially delivered into the brain. (2) spray of aerosol RNA-LNPs can be preferentially delivered into the lungs. (3) RNA-LNPs expressing GALA peptide can be preferentially delivered into the lungs. (4) RNA-LNPs expressing MAdCAM-1 D1 and D2 peptides can be preferentially delivered into the gut. (5) RNA-LNPs expressing keratinocyte targeting peptide can be extensively delivered into the skin. (6) RNA-LNPs expressing retina targeting peptide can be extensively delivered into the neural retina (photoreceptor cells) in eyes. (7) RNA-LNPs composed of 50% DOTAP (positively charged lipid) can be preferentially delivered into lung. (8) RNA-LNPs containing C12-200 can be specifically delivered into bone. (9) RNA-LNPs containing 18PA (negatively charged lipid) can be specifically delivered into the spleen. (10) Most RNA-LNPs without specific LNP design features may be preferentially delivered into the liver.

8.2.1.2 GalNAc

GalNAc is a high-affinity targeting ligand of the Asialoglycoprotein receptor (ASGPR) [146]. When GalNAc binds to ASGPR, it can enter the endosome through clathrin-mediated endocytosis. Then, ASGPR dissociates from GalNAc-siRNA conjugate and circulates back to the surface of cells. The delivery strategy based on GalNAc offers more advantages compared to LNP due to its subcutaneous administration capability and ease of scaling up. Currently, a significant number of RNA drugs in clinical trials are utilizing GalNAc technology.

8.2.1.3 Lipid-based nanoparticles (LBNPs)

Lipid-based nanoparticles (LBNPs) encompass liposomes, solid lipid nanoparticles (SLN), and nanostructured lipid carriers (NLC). These nanoparticles possess the ability to transport both hydrophobic and hydrophilic molecules, exhibit minimal or no toxicity, and prolong drug action duration [147]. The advantages of LBNPs include excellent temporal and thermal stability, high loading capacity, ease of preparation, low production costs, and suitability for large-scale industrial production due to their natural source origin. LBNPs have been extensively evaluated *in vitro* for cancer therapy as well as *in vivo,* with promising outcomes observed in certain clinical trials [148]. The incorporation of chemotherapeutic agents into lipid nanoparticles reduces the required therapeutic dosage and associated toxicity while also diminishing drug resistance. Furthermore, it enhances drug levels within tumor tissue by minimizing accumulation in healthy tissue [148].

8.2.1.4 Polymeric nanoparticles

Polymeric nanoparticles (NPs) are particles ranging in size from 1 to 1000 nm and can be loaded with active compounds entrapped within or surface-adsorbed onto the polymeric core. Polymeric NPs have garnered significant interest due to their unique properties resulting from their small size [149–151]. As drug carriers, polymeric NPs can control release, protect drugs and other biologically active molecules against environmental factors, as well as improve bioavailability and therapeutic index [151, 152].

8.2.1.5 Aptamer

Aptamer is a single-strand oligonucleotide that can bind specifically to the target molecule, which forms a specific three-dimensional structure after adaptive folding through various interaction forces such as nucleotide base complementary pairing. This binding is comparable to that of monoclonal antibodies. Compared with antibodies, aptamers have many advantages, such as a wider range of target molecules, better thermal stability, smaller molecular weight, chemical synthesis, small batch differences, easy modification, and so on. Aptamers allowed targeted delivery of RNA therapeutics through cell-specific surface markers, and once inside the cell, the nanoparticles induced lysosomal leakage that released the RNA oligonucleotides into the cytosol to achieve gene silencing [153].

8.2.1.6 Biomimetic nanomaterials

Biomimetic nanomaterials are an intersection between traditional synthetic nanomedicine and biotechnologies, offering biocompatibility, safety, and scalability of formulations with complex behaviors. The composition of biomimetic nanomaterials involves the integration of synthetic nanomaterials with simplified designs and essential biological components, which are crucial for achieving the desired effect. This approach enables enhanced biocompatibility and scalability in production [154]. Among the various biomimetic approaches investigated for enhancing the pharmacokinetics of nanoparticles (NPs), cell membranes and their constituents have emerged as a captivating opportunity [155].

8.2.2 Exosomes

Exosomes are small vesicles released by cells, typically with diameters ranging from 40 to 160 nm, which elicit minimal immune responses and possess an ability to traverse biological barriers such as cellular membranes and the blood-brain barrier [156, 157]. Exosomes contain a diverse array of biomolecules, including proteins, lipids, nucleic acids, and cell surface receptors [158, 159] and protect these molecules from degradation and facilitate intercellular communication across different tissues within the organism [82]. The challenges of utilization of exosomes as carriers for RNA delivery are how to enable large-scale production and ensure consistent quality of exosomes [160, 161]

8.2.3 Hydrogels

Macroscopic hydrogels are soft and water-swollen, which possess remarkable features such as biodegradability, tunable physiochemical properties, and injectability (**Figure 7**). Hydrogels have attracted enormous attention for use in RNA therapy recently. RNA can be loaded into hydrogels either by direct inclusion of the naked RNA or by encapsulating RNA nanocarriers. Hydrogels could improve RNA stability, reduce unnecessary loss of therapeutics during systemic delivery, mitigate undesirable off-target toxicities, and avoid the necessity of multiple doses [162].

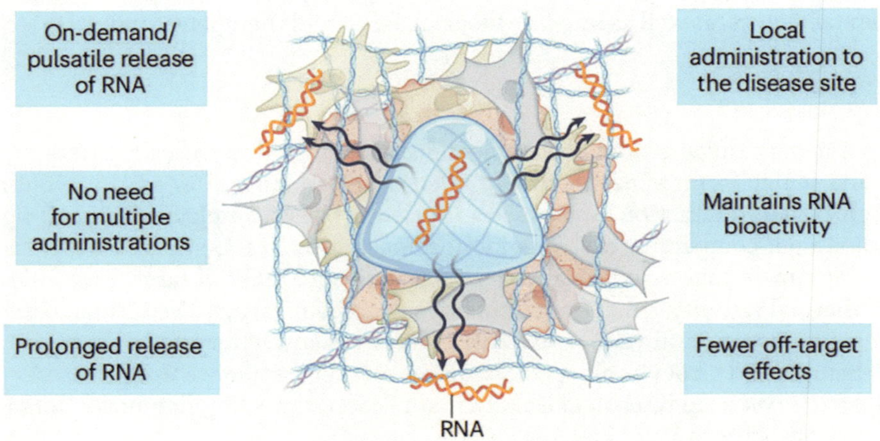

Figure 7.
Advantages of hydrogels as a platform for RNA delivery [162]. Hydrogels provide a unique strategy for local administration of RNA, overcoming some of the difficulties associated with systemic RNA delivery. They enable a localized, controlled, and sustained delivery of high levels of payloads, while maintaining RNA biological activity. Off-target effects and the need for multiple payload administrations in systemic delivery may thus be avoided.

8.2.4 Protein/peptide

A protein derived from retrovirus is found to transport RNAs outside the cell in virus-like particles (VLPs). This selective endogenous encapsidation for cellular delivery enables the delivery of exogenous mRNA cargos, such as Cre and Cas9, into cells *in vitro* without the use of non-human components [163]. As a fully human system, it may be a safer alternative to current methods, but it is still in its infancy.

8.2.5 Virus-like particles (VLPs)

VLPs possess protein coats resembling viruses, which are capable of adapting to fluctuations in PAH and temperature. Their surface can be customized with various peptides to target distinct cell types (**Figure 8**) [165].

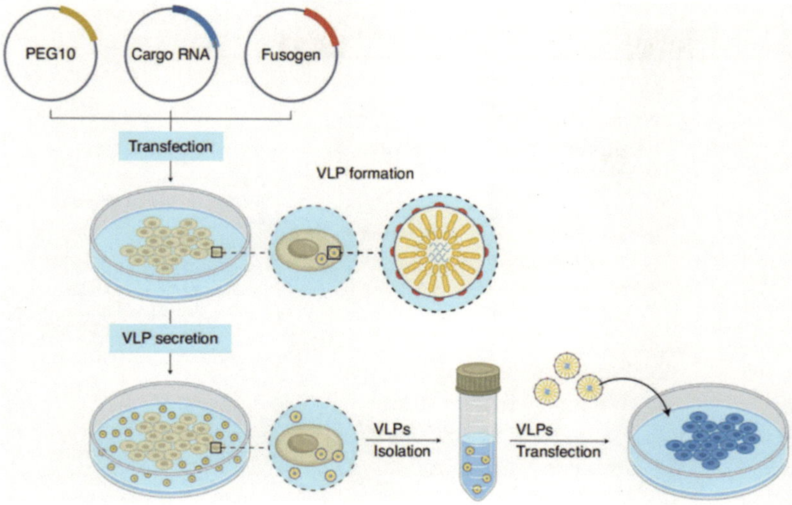

Figure 8.
Delivery of mRNA cargo by virus-like particles derived from endogenous retroelements [164]. Schematic of the "selective endogenous encapsidation for cellular delivery" (SEND) system. PEG10, cargo RNA, and fusogen vectors are transfected into cells. Inside cells, the PEG10 proteins pack the cargo mRNA and assemble into virus-like particles (VLPs) that are secreted to the growth medium in extracellular vesicles. The medium is then collected, and the VLPs are isolated using ultracentrifugation. Finally, the target cells are transfected with the VLPs. Portions of this figure were created with BioRender.com.

9. Molecular docking in RNA-based therapeutics

Molecular docking is a key tool in structural molecular biology and computer-assisted drug design. The successful docking methods effectively explore high-dimensional spaces and employ a scoring function that accurately ranks candidate dockings. Docking can be utilized for virtual screening of extensive compound libraries, ranking the outcomes, and proposing structural hypotheses on how RNAs inhibit targets, which is invaluable in lead optimization [166].

The rapidly expanding therapeutic interest in RNA-targeted drug discovery has led to an increasing demand for computational tools that can predict interactions between RNA and ligands. Virtual screening remains a crucial initial step in the design of novel drugs, particularly when only information about the targeted RNA is available. Artificial intelligence (AI) is often portrayed as a new Industrial Revolution [167]. AI finds applications in various domains, including molecular simulation for drug discovery [17]. Various methods for docking (sampling) and scoring have been developed to expedite this process, thereby enhancing our understanding of the mechanisms underlying RNA-ligand binding. Physics-based and knowledge-based approaches have demonstrated promising success in predicting both ligand-binding poses and affinities [168]. The process of molecular docking verification, see **Figure 9**.

It has been demonstrated that neglecting the flexibility of the protein (receptor) during protein-ligand docking can lead to inaccurate predictions of binding modes [169]. This issue becomes even more critical when dealing with ligand binding to RNA, which exhibits greater structural flexibility compared to proteins. RNA-Ligand DOCKing (RLDOCK) is a recently developed docking model for flexible ligands using a multiconformer approach [170, 171].

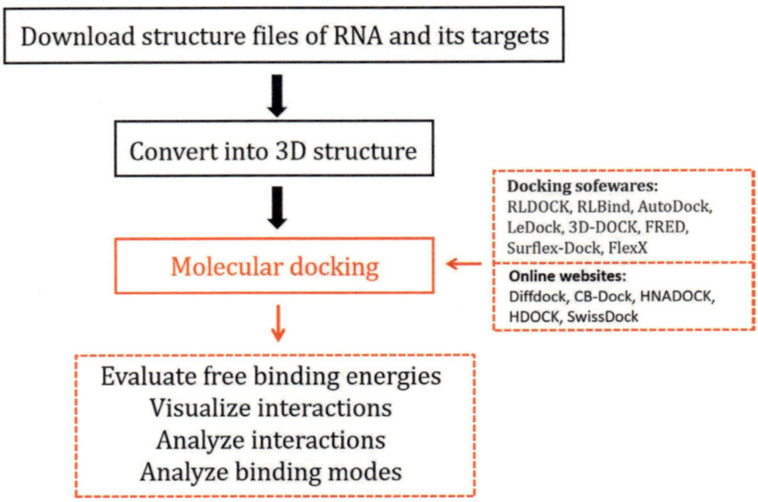

Figure 9.
The process of molecular docking verification.

Successful computational drug discovery requires the integration of three essential components: (1) a methodology for identifying druggable RNA targets, (2) a computationally efficient sampling algorithm for exploring RNA conformations, ligand conformers, and ligand binding poses, and (3) accurate scoring functions to evaluate the structures of RNA-ligand complexes and assess their binding affinity [168].

Due to the limitations of the delivery system, it is currently restricted to small RNA delivery. However, emerging evidence highlights the crucial role of lncRNAs and circRNAs in disease pathogenesis. Given their extended length, current delivery systems are inadequate for introducing these RNAs into the body. Molecular docking analysis between RNA and target molecules enables the identification of active domains and nucleotide sites that exert local key functions. Consequently, short functional RNA sequences can be designed for these regions and introduced into the body using an appropriate delivery system. Upon demonstrating efficacy at cellular and animal levels, this approach holds promise for developing vaccines or nucleic acid therapeutics suitable for clinical applications.

10. Prospects

In order to improve the targeting and efficacy of RNA therapeutics, improving RNA stability, reducing RNA-induced immunological reactions, and improving the efficiency of RNA delivery systems are the main challenges facing the development of RNA drugs in the future. The integration of ncRNA identification against conserved regions within key genomic sites associated with PAH, coupled with the development of safe nanomedicines and the production of dry powders for inhalation to enhance storage stability, positions us favorably in our pursuit of RNA therapeutics.

Using computer-aided molecular docking technology to predict the interaction and affinity of ncRNAs with their targets through 3D structure, it is helpful to construct ncRNA drugs based on the finding of effective domains of ncRNAs. Then, the constructed RNA will play therapeutic roles through suitable in vivo delivery systems. With the development of various computational tools developed for RNA-targeted drug discovery, a CASP- and D3R-like [172, 173] community-wide events with blind tests and well-curated benchmark datasets, similar to those the benchmarks widely used in the protein-ligand modeling community, would be much needed [67, 174–178].

In addition, the discovery of specific markers for various cells and specific lipid components that can target the delivery of RNA-LNPs is crucial to improving the targeting effect of RNA delivery systems. The conjugation and expression of specific peptides or aptamers on RNA-LNPs are likely to be the most reliable strategy to recognize and deliver RNA-LNPs into specific organs and/or cells. In the future, advances in basic research will make it possible to achieve the precision and efficiency of ncRNA therapies.

Moreover, it is imperative to establish a correlation between the outcomes of our *in vivo* studies and computational as well as experimental parameters. Computational approaches can be employed to accurately predict efficient ncRNA nanocarriers. Looking ahead, we are actively developing a machine-learning algorithm that will facilitate expedited clinical trials with RNA therapeutics by effectively predicting the most suitable ncRNA nanocarriers.

In conclusion, the roles of ncRNAs in PAH have been continuously clarified. Treating PAH patients with ncRNAs through appropriate delivery systems will be the trend of PAH therapy in the future. The application of molecular docking technology will help ncRNAs play a targeted therapeutic role in clinical medicine.

Author details

Yingqi Wang[1], Daling Zhu[2,3] and Hang Yu[1*]

1 Department of Basic Medicine, Harbin Medical University, Daqing, Heilongjiang Province, China

2 Central Laboratory, Harbin Medical University, Daqing, China

3 College of Pharmacy, Harbin Medical University, Daqing, China

*Address all correspondence to: yuhang310@126.com

References

[1] Galiè N, Humbert M, Vachiery JL, et al. 2015 ESC/ERS guidelines for the diagnosis and treatment of pulmonary hypertension. Kardiologia Polska. 2015;**73**(12):1127-1206

[2] Weber L, Rickli H, Haager PK, et al. Haemodynamic mechanisms and long-term prognostic impact of pulmonary hypertension in patients with severe aortic stenosis undergoing valve replacement. European Journal of Heart Failure. 2019;**21**(2):172-181

[3] Bourgeois A, Lambert C, Habbout K, et al. FOXM1 promotes pulmonary artery smooth muscle cell expansion in pulmonary arterial hypertension. Journal of Molecular Medicine (Berlin, Germany). 2018;**96**(2):223-235

[4] Bourgeois A, Omura J, Habbout K, Bonnet S, Boucherat O. Pulmonary arterial hypertension: New pathophysiological insights and emerging therapeutic targets. The International Journal of Biochemistry & Cell Biology. 2018;**104**:9-13

[5] Chelladurai P, Seeger W, Pullamsetti SS. Epigenetic mechanisms in pulmonary arterial hypertension: The need for global perspectives. European Respiratory Review. 2016;**25**(140):135-140

[6] Veith C, Schermuly RT, Brandes RP, Weissmann N. Molecular mechanisms of hypoxia-inducible factor-induced pulmonary arterial smooth muscle cell alterations in pulmonary hypertension. The Journal of Physiology. 2016;**594**(5):1167-1177

[7] Wakiyama M, Yokoyama S. MicroRNA-mediated deadenylation in a mammalian cell-free system. Methods in Molecular Biology. 2014;**1125**:341-351

[8] Xiao L, Wang J, Ju S, Cui M, Jing R. Disorders and roles of tsRNA, snoRNA, snRNA and piRNA in cancer. Journal of Medical Genetics. 2022;**59**(7):623-631

[9] Botti G, Marra L, Malzone MG, et al. LncRNA HOTAIR as prognostic circulating marker and potential therapeutic target in patients with tumor diseases. Current Drug Targets. 2017;**18**(1):27-34

[10] Di X, Jin X, Li R, Zhao M, Wang K. CircRNAs and lung cancer: Biomarkers and master regulators. Life Sciences. 2019;**220**:177-185

[11] Correia de Sousa M, Gjorgjieva M, Dolicka D, Sobolewski C, Foti M. Deciphering miRNAs' action through miRNA editing. International Journal of Molecular Sciences. 2019;**20**(24):6249

[12] Crooke ST, Witztum JL, Bennett CF, Baker BF. RNA-targeted therapeutics. Cell Metabolism. 2018;**27**(4):714-739

[13] Yin W, Rogge M. Targeting RNA: A transformative therapeutic strategy. Clinical and Translational Science. 2019;**12**(2):98-112

[14] Yu AM, Jian C, Yu AH, Tu MJ. RNA therapy: Are we using the right molecules? Pharmacology & Therapeutics. 2019;**196**:91-104

[15] Yu AM, Choi YH, Tu MJ. RNA drugs and RNA targets for small molecules: Principles, progress, and challenges. Pharmacological Reviews. 2020;**72**(4):862-898

[16] Connelly C, Moon M, Schneekloth J. The emerging role of RNA as a therapeutic target for small molecules. Cell Chemical Biology. 2016;**23**(9):1077-1090

[17] Somody JC, MacKinnon SS, Windemuth A. Structural coverage of the proteome for pharmaceutical applications. Drug Discovery Today. 2017;**22**(12):1792-1799

[18] Warner KD, Hajdin CE, Weeks KM. Principles for targeting RNA with drug-like small molecules. Nature Reviews Drug Discovery. 2018;**17**(8):547-558

[19] Tessaro F, Scapozza L. How 'protein-docking' translates into the new emerging field of docking small molecules to nucleic acids? Molecules. 2020;**25**(12):2749-2764

[20] Costales MG, Childs-Disney JL, Haniff HS, Disney MD. How we think about targeting RNA with small molecules. Journal of Medicinal Chemistry. 2020;**63**(17):8880-8900

[21] Clamp M, Fry B, Kamal M, Xie X, Cuff J, Lin MF, et al. Distinguishing protein-coding and noncoding genes in the human genome. Proceedings of the National Academy of Sciences. 2007;**104**(49):19428-19433

[22] Hopkins AL, Groom CR. The druggable genome. Nature Reviews Drug Discovery. 2002;**1**(9):727-730

[23] Dixon SJ, Stockwell BR. Identifying druggable disease-modifying gene products. Current Opinion in Chemical Biology. 2009;**13**(5):549-555

[24] Overington JP, Al-Lazikani B, Hopkins AL. How many drug targets are there? Nature Reviews Drug Discovery. 2006;**5**(12):993-996

[25] Jordan PC, Stevens SK, Deval J. Nucleosides for the treatment of respiratory RNA virus infections. Antiviral Chemistry & Chemotherapy. 2018;**26**:2040206618764483

[26] Sullenger BA, Nair S. From the RNA world to the clinic. Science. 2016;**352**(6292):1417-1420

[27] Dorsett Y, Tuschl T. siRNAs: Applications in functional genomics and potential as therapeutics. Nature Reviews. Drug Discovery. 2004;**3**(4):318-329

[28] Kifle ZD, Ayele AG, Enyew EF. Drug repurposing approach, potential drugs, and novel drug targets for COVID-19 treatment. Journal of Environmental and Public Health. 2021;**2021**:6631721

[29] Liu C, Zhou Q, Li Y, et al. Research and Development on therapeutic agents and vaccines for COVID-19 and related human coronavirus diseases. ACS Central Science. 2020;**6**(3):315-331

[30] Rothman AM, Arnold ND, Pickworth JA, et al. MicroRNA-140-5p and SMURF1 regulate pulmonary arterial hypertension. The Journal of Clinical Investigation. 2016;**126**(7):2495-2508

[31] Düzgün Z, Kayıkçıoğlu LM, Aktan Ç, et al. Decreased circulating microRNA-21 and microRNA-143 are associated to pulmonary hypertension. Turkish Journal of Medical Sciences. 2023;**53**(1):130-141

[32] Wang F, Zuroske T, Watts JK. RNA therapeutics on the rise. Nature Reviews. Drug Discovery. 2020;**19**(7):441-442

[33] Roberts TC, Langer R, Wood M. Advances in oligonucleotide drug delivery. Nature Reviews. Drug Discovery. 2020;**19**(10):673-694

[34] Kulkarni JA, Witzigmann D, Thomson SB, et al. The current landscape of nucleic acid therapeutics. Nature Nanotechnology. 2021;**16**(6):630-643

[35] Merkel OM, Librizzi D, Pfestroff A, et al. Stability of siRNA polyplexes

from poly(ethylenimine) and poly(ethylenimine)-g-poly(ethylene glycol) under *in vivo* conditions: Effects on pharmacokinetics and biodistribution measured by Fluorescence Fluctuation Spectroscopy and Single Photon Emission Computed Tomography (SPECT) imaging. Journal of Controlled Release. 2009;**138**(2):148-159

[36] Titze-de-Almeida R, David C, Titze-de-Almeida SS. The race of 10 synthetic RNAi-based drugs to the pharmaceutical market. Pharmaceutical Research. 2017;**34**(7):1339-1363

[37] Mehta A, Michler T, Merkel OM. siRNA therapeutics against respiratory viral infections-what have we learned for potential COVID-19 therapies. Advanced Healthcare Materials. 2021;**10**(7):e2001650

[38] Winter J, Jung S, Keller S, Gregory RI, Diederichs S. Many roads to maturity: microRNA biogenesis pathways and their regulation. Nature Cell Biology. 2009;**11**(3):228-234

[39] Ling H, Fabbri M, Calin GA. MicroRNAs and other non-coding RNAs as targets for anticancer drug development. Nature Reviews. Drug Discovery. 2013;**12**(11):847-865

[40] Pritchard CC, Cheng HH, Tewari M. MicroRNA profiling: Approaches and considerations. Nature Reviews. Genetics. 2012;**13**(5):358-369

[41] Gu S, Jin L, Zhang F, Sarnow P, Kay MA. Biological basis for restriction of microRNA targets to the 3′ untranslated region in mammalian mRNAs. Nature Structural & Molecular Biology. 2009;**16**(2):144-150

[42] Thenappan T, Ormiston ML, Ryan JJ, Archer SL. Pulmonary arterial hypertension: Pathogenesis and clinical management. BMJ. 2018;**360**:j5492

[43] Bertero T, Lu Y, Annis S, et al. Systems-level regulation of microRNA networks by miR-130/301 promotes pulmonary hypertension. The Journal of Clinical Investigation. 2014;**124**(8):3514-3528

[44] Chen T, Zhou G, Zhou Q, et al. Loss of microRNA-17~92 in smooth muscle cells attenuates experimental pulmonary hypertension via induction of PDZ and LIM domain 5. American Journal of Respiratory and Critical Care Medicine. 2015;**191**(6):678-692

[45] Fu J, Bai P, Chen Y, Yu T, Li F. Inhibition of miR-495 improves both vascular Remodeling and angiogenesis in pulmonary hypertension. Journal of Vascular Research. 2019;**56**(2):97-106

[46] Ma W, Qiu Z, Bai Z, et al. Inhibition of microRNA-30a alleviates vascular remodeling in pulmonary arterial hypertension. Molecular Therapy - Nucleic Acids. 2021;**26**:678-693

[47] Jiang Y, Liu H, Yu H, et al. Circular RNA Calm4 regulates hypoxia-induced pulmonary arterial smooth muscle cells Pyroptosis via the Circ-Calm4/miR-124-3p/PDCD6 Axis. Arteriosclerosis, Thrombosis, and Vascular Biology. 2021;**41**(5):1675-1693

[48] Zhang J, Li Y, Qi J, et al. Circ-calm4 serves as an miR-337-3p sponge to regulate Myo10 (myosin 10) and promote pulmonary artery smooth muscle proliferation. Hypertension. 2020;**75**(3):668-679

[49] Errington N, Iremonger J, Pickworth JA, et al. A diagnostic miRNA signature for pulmonary arterial hypertension using a consensus machine learning approach. eBioMedicine. 2021;**69**:103444

[50] Slack FJ, Chinnaiyan AM. The role of non-coding RNAs in oncology. Cell. 2019;**179**(5):1033-1055

[51] Prakash S, Dhanushkodi NR, Zayou L, et al. Cross-protection induced by highly conserved human B, CD4+, and CD8+ T cell epitopes-based coronavirus vaccine against severe infection, disease, and death caused by multiple SARS-CoV-2 variants of concern. bioRxiv. 24 May 2023. 541850 [Preprint]

[52] Hu J, Xu Y, Hao J, Wang S, Li C, Meng S. MiR-122 in hepatic function and liver diseases. Protein & Cell. 2012;**3**(5):364-371

[53] Zhang L, Liao Y, Tang L. MicroRNA-34 family: A potential tumor suppressor and therapeutic candidate in cancer. Journal of Experimental & Clinical Cancer Research. 2019;**38**(1):53

[54] Parrella P, Barbano R, Pasculli B, et al. Evaluation of microRNA-10b prognostic significance in a prospective cohort of breast cancer patients. Molecular Cancer. 2014;**13**:142

[55] Inoue J, Inazawa J. Cancer-associated miRNAs and their therapeutic potential. Journal of Human Genetics. 2021;**66**(9):937-945

[56] Chang WT, Lee WC, Lin YW, et al. Transpulmonary expression of exosomal microRNAs in idiopathic and congenital heart disease-related pulmonary arterial hypertension. Journal of the American Heart Association. 2023;**12**(23):e031435

[57] Ali MK, Schimmel K, Zhao L, et al. The role of circular RNAs in pulmonary hypertension. The European Respiratory Journal. 2022;**60**(6):2200012

[58] Yu CY, Kuo HC. The emerging roles and functions of circular RNAs and their generation. Journal of Biomedical Science. 2019;**26**(1):29

[59] Li J, Xu Q, Huang ZJ, et al. CircRNAs: A new target for the diagnosis and treatment of digestive system neoplasms. Cell Death & Disease. 2021;**12**(2):205

[60] Salmena L, Poliseno L, Tay Y, Kats L, Pandolfi PP. A ceRNA hypothesis: The Rosetta stone of a hidden RNA language. Cell. 2011;**146**(3):353-358

[61] Du WW, Yang W, Liu E, Yang Z, Dhaliwal P, Yang BB. Foxo3 circular RNA retards cell cycle progression via forming ternary complexes with p21 and CDK2. Nucleic Acids Research. 2016;**44**(6):2846-2858

[62] Legnini I, Di Timoteo G, Rossi F, et al. Circ-ZNF609 is a circular RNA that can be translated and functions in myogenesis. Molecular Cell. 2017;**66**(1):22-37.e9

[63] Conn SJ, Pillman KA, Toubia J, et al. The RNA binding protein quaking regulates formation of circRNAs. Cell. 2015;**160**(6):1125-1134

[64] Pan Z, Cai J, Lin J, et al. A novel protein encoded by circFNDC3B inhibits tumor progression and EMT through regulating snail in colon cancer. Molecular Cancer. 2020;**19**(1):71

[65] Hanan M, Soreq H, Kadener S. CircRNAs in the brain. RNA Biology. 2017;**14**(8):1028-1034

[66] Kristensen LS, Hansen TB, Venø MT, Kjems J. Circular RNAs in cancer: Opportunities and challenges in the field. Oncogene. 2018;**37**(5):555-565

[67] Li H, Li K, Lai W, et al. Comprehensive circular RNA profiles in plasma reveals that circular RNAs can be used as novel biomarkers for systemic lupus erythematosus. Clinica Chimica Acta. 2018;**480**:17-25

[68] Vo JN, Cieslik M, Zhang Y, et al. The landscape of circular RNA in cancer. Cell. 2019;**176**(4):869-881.e13

[69] Chouvarine P, Photiadis J, Cesnjevar R, et al. RNA expression profiles and regulatory networks in human right ventricular hypertrophy due to high pressure load. iScience. 2021;**24**(3):102232

[70] Guo HM, Liu ZP. Up-regulation of circRNA_0068481 promotes right ventricular hypertrophy in PAH patients via regulating miR-646/miR-570/miR-885. Journal of Cellular and Molecular Medicine. 2021;**25**(8):3735-3743

[71] Miao R, Wang Y, Wan J, et al. Microarray expression profile of circular RNAs in chronic thromboembolic pulmonary hypertension. Medicine (Baltimore). 2017;**96**(27):e7354

[72] Hong L, Ma X, Liu J, et al. Circular RNA-HIPK3 regulates human pulmonary artery endothelial cells function and vessel growth by regulating microRNA-328-3p/STAT3 axis. Pulmonary Circulation. 2021;**11**(2):20458940211000234

[73] Jin X, Xu Y, Guo M, et al. hsa_circNFXL1_009 modulates apoptosis, proliferation, migration, and potassium channel activation in pulmonary hypertension. Molecular Therapy - Nucleic Acids. 2021;**23**:1007-1019

[74] Zhang F, Jiang J, Qian H, Yan Y, Xu W. Exosomal circRNA: Emerging insights into cancer progression and clinical application potential. Journal of Hematology & Oncology. 2023;**16**(1):67

[75] Zhang W, Zhao J, Deng L, et al. INKILN is a novel long noncoding RNA promoting vascular smooth muscle inflammation via scaffolding MKL1 and USP10. Circulation. 2023;**148**(1):47-67

[76] Yue X, Lan F, Liu W. Serum exosomal circCCDC66 as a potential diagnostic and prognostic biomarker for pituitary adenomas. Frontiers in Oncology. 2023;**13**:1268778

[77] Huang Y, Su D, Ye B, et al. Expression and clinical significance of circular RNA hsa_circ_0003416 in pediatric pulmonary arterial hypertension associated with congenital heart disease. Journal of Clinical Laboratory Analysis. 2022;**36**(4):e24273

[78] Zhao L, Gu C, Gan Y, Shao L, Chen H, Zhu H. Exosome-mediated siRNA delivery to suppress postoperative breast cancer metastasis. Journal of Controlled Release. 2020;**318**:1-15

[79] Fan Y, Wang J, Jin W, et al. CircNR3C2 promotes HRD1-mediated tumor-suppressive effect via sponging miR-513a-3p in triple-negative breast cancer. Molecular Cancer. 2021;**20**(1):25

[80] Su L, Li X, Mao X, et al. Circ-Ntrk2 acts as a miR-296-5p sponge to activate the TGF-β1/p38 MAPK pathway and promote pulmonary hypertension and vascular remodelling. Respiratory Research. 2023;**24**(1):78

[81] Kristensen LS, Jakobsen T, Hager H, Kjems J. The emerging roles of circRNAs in cancer and oncology. Nature Reviews Clinical Oncology. 2022;**19**(3):188-206

[82] Zhao Q, Liu J, Deng H, et al. Targeting mitochondria-located circRNA SCAR alleviates NASH via reducing mROS output. Cell. 2020;**183**(1):76-93.e22

[83] Wesselhoeft RA, Kowalski PS, Parker-Hale FC, Huang Y, Bisaria N, Anderson DG. RNA circularization diminishes immunogenicity and can extend translation duration In vivo. Molecular Cell. 2019;**74**(3):508-520.e4

[84] Lavenniah A, Luu T, Li YP, et al. Engineered circular RNA sponges act as miRNA inhibitors to attenuate pressure

overload-induced cardiac hypertrophy. Molecular Therapy. 2020;**28**(6):1506-1517

[85] Liu X, Abraham JM, Cheng Y, et al. Synthetic circular RNA functions as a miR-21 sponge to suppress gastric carcinoma cell proliferation. Molecular Therapy - Nucleic Acids. 2018;**13**:312-321

[86] Schreiner S, Didio A, Hung LH, Bindereif A. Design and application of circular RNAs with protein-sponge function. Nucleic Acids Research. 2020;**48**(21):12326-12335

[87] Wang Z, Ma K, Cheng Y, et al. Synthetic circular multi-miR sponge simultaneously inhibits miR-21 and miR-93 in esophageal carcinoma. Laboratory Investigation. 2019;**99**(10):1442-1453

[88] Qu L, Yi Z, Shen Y, et al. Circular RNA vaccines against SARS-CoV-2 and emerging variants. Cell. 2022;**185**(10):1728-1744.e16

[89] Li H, Peng K, Yang K, et al. Circular RNA cancer vaccines drive immunity in hard-to-treat malignancies. Theranostics. 2022;**12**(14):6422-6436

[90] Ponting CP, Oliver PL, Reik W. Evolution and functions of long noncoding RNAs. Cell. 2009;**136**(4):629-641

[91] Wierzbicki AT, Haag JR, Pikaard CS. Noncoding transcription by RNA polymerase pol IVb/Pol V mediates transcriptional silencing of overlapping and adjacent genes. Cell. 2008;**135**(4):635-648

[92] Malakar P, Shilo A, Mogilevsky A, et al. Long noncoding RNA MALAT1 promotes hepatocellular carcinoma development by SRSF1 upregulation and mTOR activation. Cancer Research. 2017;**77**(5):1155-1167

[93] Rinn JL, Chang HY. Genome regulation by long noncoding RNAs. Annual Review of Biochemistry. 2012;**81**:145-166

[94] Ørom UA, Shiekhattar R. Long noncoding RNAs usher in a new era in the biology of enhancers. Cell. 2013;**154**(6):1190-1193

[95] Eptaminitaki GC, Stellas D, Bonavida B, Baritaki S. Long non-coding RNAs (lncRNAs) signaling in cancer chemoresistance: From prediction to druggability. Drug Resistance Updates. 2022;**65**:100866

[96] Omura J, Habbout K, Shimauchi T, et al. Identification of long noncoding RNA H19 as a new biomarker and therapeutic target in right ventricular failure in pulmonary arterial hypertension. Circulation. 2020;**142**(15):1464-1484

[97] Wang S, Luo Z, Yuan L, et al. tRNA-derived small RNAs: Novel insights into the pathogenesis and treatment of cardiovascular diseases. Journal of Cardiovascular Translational Research. 2023;**16**(2):300-309

[98] Wang T, Sung TC, Yu T, et al. Next-generation materials for RNA-lipid nanoparticles: Lyophilization and targeted transfection. Journal of Materials Chemistry B. 2023;**11**(23):5083-5093

[99] Wang Y, Chen D, Xie H, et al. LncRNA GAS5 suppresses TGF-β1-induced transformation of pulmonary pericytes into myofibroblasts by recruiting KDM5B and promoting H3K4me2/3 demethylation of the PDGFRα/β promoter. Molecular Medicine. 2023;**29**(1):32

[100] Yu J, Wang W, Yang J, et al. LncRNA PSR regulates vascular remodeling through encoding a novel protein Arteridin. Circulation Research. 2022;**131**(9):768-787

[101] Yang L, Liang H, Shen L, Guan Z, Meng X. LncRNA Tug1 involves in the pulmonary vascular remodeling in mice with hypoxic pulmonary hypertension via the microRNA-374c-mediated Foxc1. Life Sciences. 2019;**237**:116769

[102] Su H, Xu X, Yan C, et al. LncRNA H19 promotes the proliferation of pulmonary artery smooth muscle cells through AT1R via sponging let-7b in monocrotaline-induced pulmonary arterial hypertension. Respiratory Research. 2018;**19**(1):254

[103] Su H, Liu B, Chen H, et al. LncRNA ANRIL mediates endothelial dysfunction through BDNF downregulation in chronic kidney disease. Cell Death & Disease. 2022;**13**(7):661

[104] Zahid KR, Raza U, Chen J, Raj UJ, Gou D. Pathobiology of pulmonary artery hypertension: Role of long non-coding RNAs. Cardiovascular Research. 2020;**116**(12):1937-1947

[105] Gong Q, Hu Z, Jin Q, et al. Identification of JPX-RABEP1 pair as an immune-related biomarker and therapeutic target in pulmonary arterial hypertension by bioinformatics and experimental analyses. International Journal of Molecular Sciences. 2022;**23**(24)

[106] Zhuo Y, Zeng Q, Zhang P, Li G, Xie Q, Cheng Y. Functional polymorphism of lncRNA MALAT1 contributes to pulmonary arterial hypertension susceptibility in Chinese people. Clinical Chemistry and Laboratory Medicine. 2017;**55**(1):38-46

[107] Wang X, Li Q, He S, et al. LncRNA FENDRR with m6A RNA methylation regulates hypoxia-induced pulmonary artery endothelial cell pyroptosis by mediating DRP1 DNA methylation. Molecular Medicine. 2022;**28**(1):126

[108] Song R, Lei S, Yang S, Wu SJ. LncRNA PAXIP1-AS1 fosters the pathogenesis of pulmonary arterial hypertension via ETS1/WIPF1/RhoA axis. Journal of Cellular and Molecular Medicine. 2021;**25**(15):7321-7334

[109] Han S, Chen X, Huang L. The tumor therapeutic potential of long non-coding RNA delivery and targeting. Acta Pharmaceutica Sinica B. 2023;**13**(4):1371-1382

[110] Peng Y, Huang X, Wang H. Serum lncRNA LINC01535 as biomarker of diagnosis, prognosis, and disease progression in breast cancer. Clinical Breast Cancer. 2023;**23**(6):620-627

[111] Hsu XR, Wu JE, Wu YY, et al. Exosomal long noncoding RNA MLETA1 promotes tumor progression and metastasis by regulating the miR-186-5p/EGFR and miR-497-5p/IGF1R axes in non-small cell lung cancer. Journal of Experimental & Clinical Cancer Research. 2023;**42**(1):283

[112] Bourdoumis A, Papatsoris AG, Chrisofos M, Efstathiou E, Skolarikos A, Deliveliotis C. The novel prostate cancer antigen 3 (PCA3) biomarker. International Brazilian Journal of Urology. 2010;**36**(6):665-668; discussion 669

[113] Greco S, Zaccagnini G, Perfetti A, et al. Long noncoding RNA dysregulation in ischemic heart failure. Journal of Translational Medicine. 2016;**14**(1):183

[114] Girard A, Sachidanandam R, Hannon GJ, Carmell MA. A germline-specific class of small RNAs binds mammalian Piwi proteins. Nature. 2006;**442**(7099):199-202

[115] Aravin A, Gaidatzis D, Pfeffer S, et al. A novel class of small RNAs bind to MILI protein in mouse testes. Nature. 2006;**442**(7099):203-207

[116] Halbach R, Miesen P, Joosten J, et al. A satellite repeat-derived piRNA controls embryonic development of Aedes. Nature. 2020;**580**(7802):274-277

[117] Guo Z, Li Y, Ding SW. Small RNA-based antimicrobial immunity. Nature Reviews. Immunology. 2019;**19**(1):31-44

[118] Ma C, Zhang L, Wang X, et al. piRNA-63076 contributes to pulmonary arterial smooth muscle cell proliferation through acyl-CoA dehydrogenase. Journal of Cellular and Molecular Medicine. 2020;**24**(9):5260-5273

[119] Lipps C, Northe P, Figueiredo R, et al. Non-invasive approach for evaluation of pulmonary hypertension using extracellular vesicle-associated small non-coding RNA. Biomolecules. 2019;**9**(11):666

[120] Liu Y, Dou M, Song X, et al. The emerging role of the piRNA/piwi complex in cancer. Molecular Cancer. 2019;**18**(1):123

[121] Mai D, Ding P, Tan L, et al. PIWI-interacting RNA-54265 is oncogenic and a potential therapeutic target in colorectal adenocarcinoma. Theranostics. 2018;**8**(19):5213-5230

[122] Esteller M. Non-coding RNAs in human disease. Nature Reviews. Genetics. 2011;**12**(12):861-874

[123] Maxwell ES, Fournier MJ. The small nucleolar RNAs. Annual Review of Biochemistry. 1995;**64**:897-934

[124] Mallardo M, Poltronieri P, D'Urso OF. Non-protein coding RNA biomarkers and differential expression in cancers: A review. Journal of Experimental & Clinical Cancer Research. 2008;**27**(1):19

[125] Jaé N, Dimmeler S. Noncoding RNAs in vascular diseases. Circulation Research. 2020;**126**(9):1127-1145

[126] Al Tabosh T, Liu H, Koça D, et al. Impact of heterozygous ALK1 mutations on the transcriptomic response to BMP9 and BMP10 in endothelial cells from hereditary hemorrhagic telangiectasia and pulmonary arterial hypertension donors. Angiogenesis. 2024;**27**(2):211-227

[127] Chen Q, Yan M, Cao Z, et al. Sperm tsRNAs contribute to intergenerational inheritance of an acquired metabolic disorder. Science. 2016;**351**(6271):397-400

[128] Waldron D. Non-coding RNA: Inheritance of diet-induced metabolic changes via tsRNAs. Nature Reviews. Genetics. 2016;**17**(3):128

[129] Park J, Ahn SH, Shin MG, Kim HK, Chang S. tRNA-derived small RNAs: Novel epigenetic regulators. Cancers (Basel). 2020;**12**(10):2773

[130] Chen Y, Tang Y, Hou S, et al. Differential expression spectrum and targeted gene prediction of tRNA-derived small RNAs in idiopathic pulmonary arterial hypertension. Frontiers in Molecular Biosciences. 2023;**10**:1204740

[131] Diener C, Keller A, Meese E. Emerging concepts of miRNA therapeutics: From cells to clinic. Trends in Genetics. 2022;**38**(6):613-626

[132] Finer M, Glorioso J. A brief account of viral vectors and their promise for gene therapy. Gene Therapy. 2017;**24**(1):1-2

[133] Savenkova DA, Makarova AA, Shalik IK, Yudkin DV. miRNA pathway alteration in response to non-coding RNA delivery in viral vector-based gene therapy. International Journal of Molecular Sciences. 2022;**23**(23):14954

[134] Samulski RJ, Muzyczka N. AAV-mediated gene therapy for research and

therapeutic purposes. Annual Review of Virology. 2014;**1**(1):427-451

[135] Saraiva J, Nobre RJ, Pereira de Almeida L. Gene therapy for the CNS using AAVs: The impact of systemic delivery by AAV9. Journal of Controlled Release. 2016;**241**:94-109

[136] Verdera HC, Kuranda K, Mingozzi F. AAV vector immunogenicity in humans: A long journey to successful gene transfer. Molecular Therapy. 2020;**28**(3):723-746

[137] Bulcha JT, Wang Y, Ma H, Tai P, Gao G. Viral vector platforms within the gene therapy landscape. Signal Transduction and Targeted Therapy. 2021;**6**(1):53

[138] Ghebremedhin B. Human adenovirus: Viral pathogen with increasing importance. European Journal of Microbiology and Immunology. 2014;**4**(1):26-33

[139] Allen RJ, Byrnes AP. Interaction of adenovirus with antibodies, complement, and coagulation factors. FEBS Letters. 2019;**593**(24):3449-3460

[140] Lee CS, Bishop ES, Zhang R, et al. Adenovirus-mediated gene delivery: Potential applications for gene and cell-based therapies in the new era of personalized medicine. Genes & Diseases. 2017;**4**(2):43-63

[141] Sweeney NP, Vink CA. The impact of lentiviral vector genome size and producer cell genomic to gag-pol mRNA ratios on packaging efficiency and titre. Molecular Therapy - Methods & Clinical Development. 2021;**21**:574-584

[142] Moraes FC, Pichon C, Letourneur D, Chaubet F. miRNA delivery by nanosystems: State of the art and perspectives. Pharmaceutics. 2021;**13**(11):1901

[143] Alameh MG, Tombácz I, Bettini E, et al. Lipid nanoparticles enhance the efficacy of mRNA and protein subunit vaccines by inducing robust T follicular helper cell and humoral responses. Immunity. 2021;**54**(12):2877-2892.e7

[144] Chen Y, Li Z, Chen X, Zhang S. Long non-coding RNAs: From disease code to drug role. Acta Pharmaceutica Sinica B. 2021;**11**(2):340-354

[145] Witzigmann D, Kulkarni JA, Leung J, Chen S, Cullis PR, van der Meel R. Lipid nanoparticle technology for therapeutic gene regulation in the liver. Advanced Drug Delivery Reviews. 2020;**159**:344-363

[146] Springer AD, Dowdy SF. GalNAc-siRNA conjugates: Leading the way for delivery of RNAi therapeutics. Nucleic Acid Therapeutics. 2018;**28**(3):109-118

[147] Ozpolat B, Sood AK, Lopez-Berestein G. Liposomal siRNA nanocarriers for cancer therapy. Advanced Drug Delivery Reviews. 2014;**66**:110-116

[148] García-Pinel B, Porras-Alcalá C, Ortega-Rodríguez A, et al. Lipid-based nanoparticles: Application and recent advances in cancer treatment. Nanomaterials (Basel). 2019;**9**(4):638

[149] Cano A, Ettcheto M, Chang JH, et al. Dual-drug loaded nanoparticles of Epigallocatechin-3-gallate (EGCG)/ ascorbic acid enhance therapeutic efficacy of EGCG in a APPswe/PS1dE9 Alzheimer's disease mice model. Journal of Controlled Release. 2019;**301**:62-75

[150] Cano A, Sánchez-López E, Ettcheto M, et al. Current advances in the development of novel polymeric nanoparticles for the treatment of neurodegenerative diseases. Nanomedicine (London, England). 2020;**15**(12):1239-1261

[151] Soppimath KS, Aminabhavi TM, Kulkarni AR, Rudzinski WE. Biodegradable polymeric nanoparticles as drug delivery devices. Journal of Controlled Release. 2001;**70**(1-2):1-20

[152] Owens DE 3rd, Peppas NA. Opsonization, biodistribution, and pharmacokinetics of polymeric nanoparticles. International Journal of Pharmaceutics. 2006;**307**(1):93-102

[153] Tanno T, Zhang P, Bailey C, et al. A novel aptamer-based small RNA delivery platform and its application to cancer therapy. Genes & Diseases. 2023;**10**(3):1075-1089

[154] Rampado R, Caliceti P, Agostini M. Latest advances in biomimetic cell membrane-coated and membrane-derived Nanovectors for biomedical applications. Nanomaterials (Basel). 2022;**12**(9):1543

[155] Fang RH, Kroll AV, Gao W, Zhang L. Cell membrane coating nanotechnology. Advanced Materials. 2018;**30**(23):e1706759

[156] Cui J, Wang X, Li J, et al. Immune exosomes loading self-assembled nanomicelles traverse the blood-brain barrier for chemo-immunotherapy against glioblastoma. ACS Nano. 2023 [Online ahead of print]

[157] Myo AC, Kobayashi Y, Niki Y, Kamimoto H, Moriyama K. Exosome-mediated small interfering RNA delivery inhibits aberrant osteoblast differentiation in apert syndrome model mice. Archives of Oral Biology. 2023;**153**:105753

[158] Arya SB, Collie SP, Parent CA. The ins-and-outs of exosome biogenesis, secretion, and internalization. Trends in Cell Biology. 2024;**34**(2):90-108

[159] Kalluri R, LeBleu VS. The biology, function, and biomedical applications of exosomes. Science. 2020;**367**(6478):eaau6977

[160] Darband SG, Mirza-Aghazadeh-Attari M, Kaviani M, et al. Exosomes: Natural nanoparticles as bio shuttles for RNAi delivery. Journal of Controlled Release. 2018;**289**:158-170

[161] Li X, Yu Q, Zhao R, et al. Designer exosomes for targeted delivery of a novel therapeutic cargo to enhance sorafenib-mediated ferroptosis in hepatocellular carcinoma. Frontiers in Oncology. 2022;**12**:898156

[162] Zhong R, Talebian S, Mendes BB, et al. Hydrogels for RNA delivery. Nature Materials. 2023;**22**(7):818-831

[163] Segel M, Lash B, Song J, et al. Mammalian retrovirus-like protein PEG10 packages its own mRNA and can be pseudotyped for mRNA delivery. Science. 2021;**373**(6557):882-889

[164] Gutkin A, Rosenblum D, Peer D. RNA delivery with a human virus-like particle. Nature Biotechnology. 2021;**39**(12):1514-1515

[165] Mohsen MO, Gomes AC, Vogel M, Bachmann MF. Interaction of viral capsid-derived virus-like particles (VLPs) with the innate immune system. Vaccines (Basel). 2018;**6**(3):37

[166] Morris GM, Lim-Wilby M. Molecular docking. Methods in Molecular Biology. 2008;**443**:365-382

[167] Ezkurdia I, Juan D, Rodriguez JM, et al. Multiple evidence strands suggest that there may be as few as 19, 000 human protein-coding genes. Human Molecular Genetics. 2014;**23**(22):5866-5878

[168] Zhou Y, Jiang Y, Chen SJ. RNA-ligand molecular docking: Advances

and challenges. Wiley Interdisciplinary Reviews: Computational Molecular Science. 2022;**12**(3):e1571

[169] Totrov M, Abagyan R. Flexible ligand docking to multiple receptor conformations: A practical alternative. Current Opinion in Structural Biology. 2008;**18**(2):178-184

[170] Jiang Y, Chen SJ. RLDOCK method for predicting RNA-small molecule binding modes. Methods. 2022;**197**:97-105

[171] Sun LZ, Jiang Y, Zhou Y, Chen SJ. RLDOCK: A new method for predicting RNA-ligand interactions. Journal of Chemical Theory and Computation. 2020;**16**(11):7173-7183

[172] Gathiaka S, Liu S, Chiu M, et al. D3R grand challenge 2015: Evaluation of protein-ligand pose and affinity predictions. Journal of Computer-Aided Molecular Design. 2016;**30**(9):651-668

[173] Moult J, Pedersen JT, Judson R, Fidelis K. A large-scale experiment to assess protein structure prediction methods. Proteins. 1995;**23**(3):ii-v

[174] Huang N, Shoichet BK, Irwin JJ. Benchmarking sets for molecular docking. Journal of Medicinal Chemistry. 2006;**49**(23):6789-6801

[175] Li Y, Han L, Liu Z, Wang R. Comparative assessment of scoring functions on an updated benchmark: 2. Evaluation methods and general results. Journal of Chemical Information and Modeling. 2014;**54**(6):1717-1736

[176] Li Y, Su M, Liu Z, et al. Assessing protein-ligand interaction scoring functions with the CASF-2013 benchmark. Nature Protocols. 2018;**13**(4):666-680

[177] Mysinger MM, Carchia M, Irwin JJ, Shoichet BK. Directory of useful decoys, enhanced (DUD-E): Better ligands and decoys for better benchmarking. Journal of Medicinal Chemistry. 2012;**55**(14):6582-6594

[178] Su M, Yang Q, Du Y, et al. Comparative assessment of scoring functions: The CASF-2016 update. Journal of Chemical Information and Modeling. 2019;**59**(2):895-913